MODERN
PROBABILITY
THEORY

Harper's Series in Modern Mathematics
I. N. Herstein and Gian-Carlo Rota, Editors

MODERN PROBABILITY THEORY

RUI ZONG YEH

University of Hawaii

HARPER & ROW, PUBLISHERS
New York   Evanston   San Francisco   London

For Suchu and our children
Emerald and Elm

CONTENTS

Interdependence of Sections

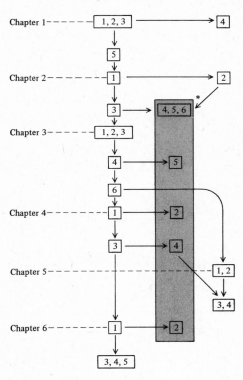

* These are relatively complicated sections.

PREFACE

This book presents as a part of modern mathematics that portion of probability theory commonly taught to juniors and seniors in most universities. Emphasis is placed on the transition from the real world to the world of mathematical models so that the student will know which aspects of the real world are being modeled. A mathematical theory is essentially a conglomeration of mathematical models, each of which ideally stipulates certain basic features of the real world, and within which deductive reasonings are carried out to obtain results that are empirically anticipated or, better yet, results that are totally unexpected.

A mathematical model, and in particular a probability model, is a complex of mathematical objects such as sets, mappings, and their numerous ramifications. Naturally, a mathematical model can be quite complicated, especially when the aspects of reality being modeled are complicated. This book will not go into the more complicated probability models generally known as random functions or stochastic processes; however, it will pursue the simple model of probability spaces to obtain important results. This is done in the first two chapters, partly as a demonstration of the usefulness of mathematical models. The last three sections of Chapter 2 on special probability spaces (Bernoulli, Markov, and Poisson) should stimulate the student to inquire about other probability models, and this, of course, is a good motivation for the study of stochastic processes.

What stopped us from going further into other probability models at the end of Chapter 2, aside from lack of space, is the lack of mathematical parts needed for further model building, and this takes us directly to an important mathematical object, the random variable, introduced in Chapter 3, and studied through Chapters 4 and 5. Random variables are indispensable not only for probabilists but also for statisticians. Statisticians generally regard random quantities observed in sampling as random variables. It is only natural, then, that probability theory should be regarded by statisticians as a necessary theoretical basis upon which to build their own statistical theory. A few statistically oriented topics are included in Chapter 5 with this in mind; some readers may want to skip these in order to get to Chapter 6 quickly.

In Chapter 6 simple mathematical models are constructed from random variables by merely adding a large number of them. Such models have as their counterpart in the real world random quantities that are actually sum totals of many small random quantities. Investigations of these models lead to the conclusion that these random quantities must have some universal features about them. These features are expressly stated as central limit theorems and laws of large numbers. These results are among the great historic triumphs of probability theory, especially since some of them are quite unexpected from an empirical point of view. Furthermore, central limit theorems proved to be a convenient starting point for statistical theory. A short course leading to these theorems may be planned with the diagram, Interdependence of Sections, that follows the contents.

For an average student this book may contain enough material to engage his attention for about a year. However, various short courses can be easily worked out by individual instructors. Some education majors may find Chapter 1 and part of Chapter 2 of use by themselves. Taking a shortest route, one can reach Chapter 6 in one semester.

The prerequisite for reading the entire book is an elementary course in calculus including multiple integrals. Chapters 1 and 2 can be read without the knowledge of calculus if the derivation of the Poisson probability vector is omitted. Chapter 3 can be an excellent review of basic theorems in calculus, while Chapter 4 introduces the Lebesgue approach to integration in about the most natural context possible. Although several measure-theoretic results are quoted in the book, these are made to appear in the right context so that the student may feel no resistance to accepting them. It is hoped that through exposures to these results (which are usually taught in a real analysis course), the student may learn to appreciate measure theory as a welcome tool for mastering the probability theory. However, if a student dislikes measure-theoretic material, he may skip the "contaminated" portion and still find the book worth reading.

This book is an outgrowth of teaching the subject for many years to a mixed audience of mathematics majors, engineers, education majors, economists, business majors, and others. Consequently, only the essential theoretical framework of probability is developed; also, since an average intelligent reader is often dismayed by obtrusive definitions and seemingly arbitrary mathematical assumptions, rigorous statements are usually preceded by short informal discussions.

R. Z. Yeh

MODERN
PROBABILITY
THEORY

Chapter 1
NAIVE
PROBABILITY
MODELS

The purpose of this chapter is to introduce the most basic type of probability model, known as probability spaces. Among these the simplest are those called finite probability spaces appearing in Section 1.3. The finiteness condition is discarded in Section 1.5 to allow the general probability spaces, for which some important special cases are considered in Chapter 2.

Section 1.1 (Outcomes and Sample Sets) and Section 1.2 (Events and Their Probabilities) are preparations for Section 1.3 (The Simplest Probability Model), whereas Section 1.4 (The Classical Formulas) is essentially a computational footnote to Section 1.3. Section 1.5 (General Probability Spaces) is, of course, the most important in the chapter.

The word "naive" in the title is used in the sense that the probability spaces considered in the chapter do not require *conditional probability*, an important concept to be brought up in Chapter 2.

1.1 OUTCOMES AND SAMPLE SETS
Probability theory is primarily concerned with building mathematical models in which events in real world are represented and their probabilities determined. Representations of events associated with random phenomena (or experiments) such as coin tossing, dice throwing, automobile accidents, and motions of small particles require first of all consideration of all possible outcomes associated with a particular random phenomenon. We shall consider in this section these outcomes collectively. In Section 1.2 we shall take up the events arising from these outcomes and think about the probabilities of these events. As we proceed, the ingredients of our mathematical models become increasingly complex. In this section we need little more than the elementary ideas of *sets* and their *elements*.

1

Let us begin by considering a simple experiment of throwing a possibly loaded die. If we represent the familiar six outcomes by I, II, III, IV, V, VI, their totality will constitute a set

$$S = \{\text{I, II, III, IV, V, VI}\}$$

which we might call a *sample set*. The elements of the sample set, generically denoted by ω (Greek omega), are then regarded as representing all possible outcomes associated with the random phenomenon in question. To take another example, if a dart is thrown at a tree, ω may represent any point on the tree, and we arrive at the sample set

$$S = \{\omega : \omega \text{ represents a point on the tree}\}$$

The braces $\{\dots\}$ should be read as "the set of all ..." and the colon inside the braces should be read as "such that." Also, if we should write

$$\omega \in S$$

this should be read as "ω is an element (or member) of S."

Different sample sets may be associated with a given experiment depending on the observer's point of view. Consider for instance the following example.

Example 1 An experiment consists of tossing a penny and a dime simultaneously. If the experimenter observes *heads* or *tails* of both coins, we have the sample set

$$S_1 = \{(h, h), (h, t), (t, h), (t, t)\}$$

If the experimenter observes instead the distance between the two coins, we have the sample set

$$S_2 = \{x : x \geqq 0\}$$

The generic symbol x instead of ω is used since it is customary to use x to represent real numbers.

Although different sample sets may be associated with a given random phenomenon according to observers' points of interest, it is often possible to consider one common sample set which is sufficiently detailed to satisfy many different observers. We illustrate this by the following example.

Example 2 At a toll-collecting station on an expressway an officer collects \$3 from each passing car from 12 noon to 6 P.M. daily. Associated with this particular random phenomenon repeated daily one may wish to observe the number of cars passed, or the amount of toll collected, or the exact moment of arrival of the first car, or the time lag between the first and the second cars, and so on. However, instead of making several fragmented observations one can unbiasedly take note of practically everything that is happening. If for instance the exact arrival time of each passing car is marked daily on the time interval [0, 6], the resulting sample set will consist of all finite monotone increasing sequences of real numbers between 0 and 6. For example the sequence

$$\omega = \langle 1.5, 2.75, 3, 5.1 \rangle$$

represents the outcome that on a particular day the first car arrived at 1:30 P.M., the second car arrived at 2:45 P.M., and so on. From this sequence it is easy to see that on this particular day the number of cars passed by is 4, the amount of toll collected is $12, the exact arrival time of the first car is 1:30 p.m., the time lag between the first and the second car is 1 hour and 15 minutes, and so forth. The sample set may be represented by

$$S = \{\langle t_1, t_2, \ldots, t_n \rangle : 0 \leqq t_1 < t_2 < \cdots < t_n \leqq 6, n \geqq 0\}$$

Going back to the dice-throwing experiment, let us consider next the relative likelihoods of outcomes represented in the sample set. Since we are assuming that we do not know enough about the die, we can only proceed empirically. Suppose, after throwing the die a hundred times, we observe the following frequencies of occurrence of the six outcomes:

Outcomes	I	II	III	IV	V	VI
Frequencies	25	54	15	4	2	0
Frequency ratios	0.25	0.54	0.15	0.04	0.02	0.00

We can tentatively make the following conclusions: The outcome II is the most likely of all, while the outcome VI is the least likely, and so on. If, as we increase the number of throws, we observe that the six frequency ratios remain more or less the same, then we might say that a certain *statistical regularity* pertaining to the likelihood of occurrences of the outcomes is in evidence, and use this statistical regularity as the empirical basis for assigning *mathematical probability* to each outcome in the sample set. Probability, as we understand it, is thus an abstraction of frequency ratio, and as such it is allowed to retain the essential attributes of frequency ratios. For example, frequency ratios are nonnegative numbers totaling 1, therefore we shall define *probabilities* of outcomes to be also nonnegative numbers totaling 1.

As for the values of six numbers representing the probabilities of the six outcomes, there does not seem to be any unique way of choosing them since the experiment may be terminated at any time or prolonged to any length. If after 100 throws we set

$$p(I) = 0.25, p(II) = 0.54, \ldots$$

this is no declaration of "absolute" truth; it is merely a reasonable choice based on the data available. Conceivably another experimenter might assign, say $p(I) = 0.251, p(II) = 0.551$, and so on. In general, construction of a mathematical model involves some subjective judgment, but once a model is chosen we can leave the real world in order to proceed objectively until some significant deductions are made within the model.

We conclude this section with the following remarks: (1) If we assign a probability value of 0 to an outcome, we are regarding that outcome as *almost impossible*; if we assign a probability value of 1 to an outcome, we are regarding that outcome as *almost certain*. If a dart is thrown at a circular board from a distance, we should regard it as almost impossible for the dart to land at the exact center of the board; in other words we should regard it as almost certain that the dart will miss the exact center. (2) If we assign equal probability values to all outcomes, we are regarding all outcomes as *equally likely*. Equal likelihood of outcomes is our way of looking at things; in reality there may be no absolutely reliable test to ascertain equal likelihood of outcomes. Also, if two outcomes are very nearly equally likely, no harm is done in assuming that they are equally likely. If a coin comes out of the United States mint, it is reasonable to assign equal probabilities to heads and tails. (3) The word "probability" is often used in a sense different from ours, as for example in "the probability that the United States stays neutral in the next world war is 0.01." Such a statement only expresses an individual's degree of conviction and has nothing to do with our kind of probability based entirely on repeatability of experiment.

1.1 PROBLEMS

1. Describe the sample set by listing all possible outcomes for each of the following experiments.
   a. Three dice are rolled and the highest number of dots is recorded.
   b. Four balls are drawn simultaneously from an urn containing many red, white, and blue balls, and their colors are observed.
   c. Five cards are drawn one after another without replacement, and the number of aces is counted.

2. Describe a sample set by drawing a sketch if convenient for each of the following:
   a. A coin is tossed until it falls heads.
   b. A man stands in the desert until it starts to rain.
   c. An executive arrives at his office at any time between 8 A.M. and 12 noon while his secretary arrives at any time between 7 and 8 A.M.
   d. A stick of 1 ft length is broken into three pieces at random.
   e. A teapot weighing 1 lb is dropped on the floor, and all broken pieces are weighed in the order of their magnitudes.

3. In the experiment of drawing 2 cards from a deck of 52 playing cards, if the observer is interested only in whether the two cards have the same color, then a sample set of two elements suffices since the colors are either *different* or *identical*. Can you think of other points of interest that an observer may take that might lead to a sample set of three elements, or four elements?

4. Thumbtacks fall heads or tails like a coin. Toss a thumbtack 10 times, 50 times, and so on. What statistical regularity do you observe? Assign probabilities to the two outcomes. Deform the thumbtack and perform the same experiment. If the stem of the thumbtack is extremely short, what probabilities would you assign to heads and tails?

5. Ask as many of your friends as possible to each pick out at random an integer from 1 to 10. Which integer is picked most often? What probability will you assign to it?

6. Think about the possibility (or impossibility) of assigning probabilities to the following events.
   a. There will be more wars in Asia than in Europe in the next decade.
   b. The Dow-Jones Industrial Average will go up by more than 10 points tomorrow.
   c. This Sunday it will rain in New York's Central Park.
   d. The next childbirth in San Francisco will be quintuplets.

1.2 EVENTS AND THEIR PROBABILITIES

Historically, the development of probability theory has its origin in the so-called games of chance. A gambler may roll a die and bet on an individual outcome, but more often than not he will bet on an *event*. For example, he may bet on the event of "odd dots"; that is, he wins whenever any one of the three outcomes I, III, V occurs. Whether it is a I, a III, or a V is of no importance to him. The event he is interested in consists of three outcomes. We may express this by writing

$$E = \{I, III, V\}$$

Other events we might consider are that of "even dots," or that of "one or six dots." Thus,

$$E_1 = \{I, III, V\}$$
$$E_2 = \{II, IV, VI\}$$
$$E_3 = \{I, VI\} \quad \text{and so on}$$

Although an event may be described verbally, it can also be specified by indicating the outcomes involved. These outcomes taken together constitute a portion of the sample set. Therefore, an event is naturally represented by a subset of the sample set.

An event is said to occur if and only if any one of its constituent outcomes occurs. However, the occurrence of an outcome may signify occurrences of several events. For example, the occurrence of the outcome VI may be considered as occurrence of E_2 or E_3 above. If an event A is a *subevent* of another event B (denoted $A \subset B$) in the sense that every outcome in A is also an outcome in B, then whenever A occurs B occurs also. In this case, it is said that the event A *implies* the event B.

Given two events A and B within a sample set S, by the *union* event $A \cup B$ of A and B we mean the event consisting of all outcomes which are either in A or in

B or possibly both; by the *intersection* event $A \cap B$ or AB we mean the event consisting of all outcomes which are both in *A* and in *B*. Two events are said to be *mutually exclusive* if they have no outcomes in common. In this case the occurrence of one event automatically precludes that of the other. Given an event *A*, by its *complement* event A^c or A' or $S - A$ we mean the event consisting of all outcomes in the sample set *S* which are not in *A*. Clearly if *B* is the complement event of *A*, then *A* is the complement event of *B*. If *A* and *B* are mutually complementary, then of course they are mutually exclusive. The two complementary events that play special roles are the *sure* event *S* consisting of all outcomes and the *impossible* event ϕ consisting of no outcomes. For example, if seven dice are tossed, the event described by "all seven faces are different" is the impossible event represented by the empty subset ϕ of the sample set *S*. Finally if an event consists of a single outcome, such an event is called a singleton event or an *elementary* event. For example, $E_4 = \{IV\}$ is an elementary event consisting of the single outcome IV.

Example 1 If we toss a coin three times, counting this as a single performance of the experiment, the event that the first toss is a tail is represented by the subset

$$\{(t, h, h), (t, h, t), (t, t, h), (t, t, t)\}$$

of the sample set, which consists of eight such triples.

Example 2 Depending on the natures of experiments and the corresponding background sample sets, events with identical verbal descriptions may have different representations in terms of outcomes. In tossing several dice simultaneously the event of "all faces alike" consists of ever smaller portions of the sample set as the number of dice is increased.

Relations that may exist among events are often visualized in terms of drawings. For instance, if *A* is a subevent of *B*, or if *A* and *B* have some outcomes in common, we may express these by the drawings in Fig. 1.2.1.

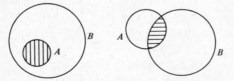

Figure 1.2.1

Such drawings suggestive of interrelations among sets are commonly known as *Venn diagrams*. They help not only in exhibiting relative positions of events within the sample set but also in discovering relations that may exist among events. We illustrate this by the following two examples.

Example 3 Use the Venn diagram (Fig. 1.2.2) to check that the events *A*, *B*, and *C* satisfy the following equations. Remember that A' designates the complement of *A* and *AB* designates the intersection of *A* and *B*, and so on.

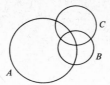

Figure 1.2.2

(i) $(A \cup B)' = A'B'$ (De Morgan's law I)
(ii) $(AB)' = A' \cup B'$ (De Morgan's law II)
(iii) $A(B \cup C) = AB \cup AC$ (distributive law I)
(iv) $A \cup (BC) = (A \cup B)(A \cup C)$ (distributive law II)

Example 4 Five decimal digits are drawn at random (typical outcomes: 73744, 09999, 68521). Let E_i be the event that some digit appears exactly i times (thus 73744 belongs to E_1 as well as E_2). Use Venn diagram to exhibit the relative positions of $E_0, E_1, \ldots, E_6$. Work it out on a piece of paper before looking at the following solution.

Solution We suggest the whole sample set S by, say, a rectangle (see Fig. 1.2.3).

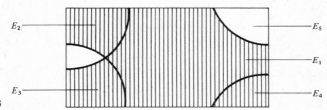

Figure 1.2.3

Since $E_6 = \phi$ and since an empty set is usually not drawn in a Venn diagram, we shall not make any drawing for E_6; on the other hand we will merely point out that $E_0 = S$. Filling in $E_5, E_4, \ldots$ roughly in that order, using the following information (the reader may check for himself), we arrive at the Venn diagram, in which the shaded area is E_1;

$$E_4 E_5 = \phi \qquad\qquad E_4 \subset E_1$$
$$E_3 E_5 = E_3 E_4 = \phi \qquad E_3 \subset E_1 \cup E_2$$
$$E_2 E_5 = E_2 E_4 = \phi \qquad E_2 \subset E_1 \cup E_3$$

Just as we considered earlier the likelihood of an outcome, so we now consider the likelihood of an event. The notion of likelihood of an outcome will be subsumed by that of likelihood of an elementary event. The determination of $P(E)$, the probability of event E, creates no difficulty since an event should be regarded to be as likely as the outcomes in the event are. In the experiment of throwing a die the frequency of occurrence of an event, say $E = \{I, III, V\}$, is clearly equal to the sum of the frequencies of occurrence of outcomes I, III, and V, so that the frequency ratio $f_n(E)$ of E corresponding to n trials of experiment is equal to $f_n(I) + f_n(III) + f_n(V)$. In keeping with the fact that the frequency ratios $f_n(I)$,

$f_n(\text{III})$, and $f_n(\text{V})$ are taken as probabilities of I, III, and V, we shall take the sum of these frequency ratios as the probability of E. Thus the probability of E should be defined as

$$P(E) = p(\text{I}) + p(\text{III}) + p(\text{V})$$

In general, we shall take the sum of probabilities of outcomes forming a event as the probability of that event. Viewed mathematically, probability P is a *set function* that assigns a nonnegative number $P(E)$ to each subset E of the sample set S in such a way that certain statements about the numbers $P(E)$ can be made. These are, and the reader can verify them in Problem 1.2.8,

(i) $0 \leq P(E) \leq 1$ for any $E \subset S$, and $P(S) = 1$.

(ii) If A and B are mutually exclusive events, then

$$P(A \cup B) = P(A) + P(B)$$

(iii) If A^c is the complement of A, then

$$P(A^c) = 1 - P(A)$$

(iv) If $A \subset B$, then $P(A) \leq P(B)$.

(v) If A and B are two events not necessarily mutually exclusive, then

$$P(A \cup B) = P(A) + P(B) - P(AB)$$

These five properties of P are useful in calculations of $P(E)$. We illustrate this by the following simple example.

Example 5 After tossing half a dozen coins simultaneously many times it is agreed that the probability of "all heads" shall be 0.088 and that of "all tails" 0.002. What then must be the probability of "heads and tails mixed"?

Solution Let H, T, M be the three events in question. Then

$$(H \cup T) \cup M = S \qquad \text{and} \qquad (H \cup T)M = \phi$$

Hence by (iii)

$$P(M) = 1 - P(H \cup T)$$

But by (ii)

$$P(H \cup T) = P(H) + P(T) = 0.088 + 0.002 = 0.09$$

Consequently

$$P(M) = 1 - 0.09 = 0.91$$

The five properties of P we have considered thus far are by themselves not sufficient for calculations of probabilities of events. For instance, how do we calculate $P(AB)$ from $P(A)$ and $P(B)$? We shall see in the next chapter that under certain conditions $P(AB) = P(A)\,P(B)$. For the time being, the calculation of

$P(AB)$ must be done without recourse to any such convenient formula. In the following example we determine $P(AB)$ in a way consistent with the idea that probability is an abstraction of frequency ratios.

Example 6 After experimenting separately with two loaded dice, it is agreed that these two dice can be described by two separate probability "distributions" as follows:

$$p_1(\text{I}) = 0.25 \qquad p_1(\text{IV}) = 0.04$$
$$p_1(\text{II}) = 0.54 \qquad p_1(\text{V}) = 0.02$$
$$p_1(\text{III}) = 0.15 \qquad p_1(\text{VI}) = 0.00$$

for the first die, and

$$p_2(\text{I}) = p_2(\text{II}) = p_2(\text{III}) = p_2(\text{IV}) = 0.15$$
$$p_2(\text{V}) = p_2(\text{VI}) = 0.20$$

for the second die. If the two dice are thrown together and A is the event that the sum of the two faces is odd and B is the event that at least one face is an ace, determine the probability of the event AB.

Solution The event AB may be represented by

$$\{(1, 2), (1, 4), (1, 6), (2, 1), (4, 1), (6, 1)\}$$

Hence the probability of AB can be determined by the probabilities of these six outcomes. Now the probability of the outcome $(1, 2)$, say, may be determined from $p_1(\text{I})$ and $p_2(\text{II})$ as follows: If we toss the pair of dice a large number of times, we expect the first die to give an ace about 25% of the time since $p_1(\text{I}) = 0.25$; also, while the second die may give any of the six faces during these 25% of the tosses, we expect it to give a 2 about 15% of the time since $p_2(\text{II}) = 0.15$. Thus we may expect the frequency ratio of the outcome $(1, 2)$ in actual repeated tosses of the dice to be somewhere near $0.25 \times 0.15 = 0.0375$. Likewise we may estimate frequency ratios for other outcomes, (i, j) for $1 \leq i \leq 6$, $1 \leq j \leq 6$. Since these frequency ratios clearly add up to 1, we can reasonably accept them as probabilities $p(i, j)$. Now therefore,

$$P(AB) = p(1, 2) + p(1, 4) + \cdots + p(6, 1)$$
$$= (0.25)(0.15) + (0.25)(0.15) + \cdots + (0.00)(0.15)$$
$$= 0.212$$

1.2 PROBLEMS

1. In the experiment of repeatedly tossing a coin until it falls heads, represent the event that the coin falls heads within the first three tosses as a subset of an appropriate set.

2. In the experiment of breaking a 1-ft stick into three pieces at random, represent the event that the left-end piece is at least 6 in. as a subset of the sample set. In the same sample set represent the event that the left-end piece is at least 6 in. while the right-end piece is at most 3 in.

3. In the experiment of dropping a teapot of 1-lb weight, represent the event that the third heaviest broken piece weighs less than 0.1 lb. Compare this event with the event that the fifth heaviest broken piece weighs less than 0.1 lb. Does either event imply the other?

4. In a certain experiment whenever the event A fails to occur the event B occurs and whenever the event B fails to occur the event A occurs. Does this necessarily mean that A and B are complementary events?

5. The occurrence of any outcome in an experiment may always be interpreted as the occurrence of any event containing this particular outcome. For the experiment of tossing two coins, write down an appropriate sample set and count the number of events that occur when both coins fall heads.

6. In the experiment of tossing a coin three times, in what way is the event E that both the first and the second tosses are tails related to the event A that the first toss is a tail and the event B that the second toss is a tail? In what way is the event D that either the first or the second (but not both in this particular situation) toss is a tail related to the events just considered?

7. A deck of 10 cards includes a joker. A card is drawn at random from the deck three times, each time after replacing the previously drawn. Let E be the event that the joker is drawn at the first draw or the third draw. Calculate the number of outcomes in E by first calculating the number of outcomes in E^c.

8. Let $S = \{\omega_i\}_{i=1}^n$ be the sample set of a given random experiment and p_i be the probability value assigned to the outcome ω_i. Thus $\{p_i\}_{i=1}^n$ is a set of nonnegative real numbers totaling 1. If we let P assign to each subset E of S the following real number

$$P(E) = \sum_i p_i \quad \text{(summation over such } i \text{ that } \omega_i \in E)$$

show that P satisfies, among other properties, the following:

$$P(A \cup B) = P(A) + P(B) - P(AB)$$

What other properties does P satisfy?

9. Two ordinary dice are rolled. Let E_i be the event that the sum of the two faces is equal to i. Which event among the E_i is the most likely? The least likely? It can be shown (cf. Problem 3.6.4) that no matter how we load the two dice it is not possible to make the E_i's equally likely.

10. In tossing a coin several times if either head or tail repeats itself uninterruptedly for a maximum of k times, we say that a *run of size k* has occurred. If a fair coin is tossed three times, what is the probability of a run of size 2?

11. Two sets (events) are *identical* if and only if every element (outcome) in one set is also an element in the other. That is, $X = Y$ if and only if $X \subset Y$ and $Y \subset X$. Give formal proofs to a few of the following identities:

    a. $(A \cup B) \cup C = A \cup (B \cup C)$ (associative law I).

    b. $(AB)C = A(BC)$ (associative law II).

    c. $A \cup B = B \cup A$ (commutative law I).

    d. $AB = BA$ (commutative law II).

    e. $A \cup B = B$ if and only if $A \subset B$.

    f. $AB = B$ if and only if $A \supset B$.

12. Given n sets $A_1, A_2, \ldots, A_n$ how would you define their union set $A_1 \cup A_2 \cup \cdots \cup A_n$ or $\bigcup_{i=1}^{n} A_i$ and their intersection set $A_1 A_2 \cdots A_n$ or $\bigcap_{i=1}^{n} A_i$?

13. Prove $(A \cup B)' = A' \cap B'$. Use this result to deduce $(A \cap B)' = A' \cup B'$. Extend these results to obtain the so-called De Morgan formulas:

$$(A_1 \cup A_2 \cup \cdots \cup A_n)' = A_1' A_2' \cdots A_n'$$

$$(A_1 A_2 \cdots A_n)' = A_1' \cup A_2' \cup \cdots \cup A_n'$$

14. If $A - B$ consists of all outcomes that are in A but not in B, is it true that $P(A - B) = P(A) - P(B)$? What if B is a subevent of A?

15. Events $E_1, E_2, \ldots, E_n$ are said to form a disjoint family if they are pairwise mutually exclusive. By means of a Venn diagram show that if $E_1 E_2 \cdots E_n = \phi$ this does not necessarily guarantee that $E_1, E_2, \ldots, E_n$ form a disjoint family.

16. A disjoint family of events $E_1, E_2, \ldots, E_n$ is said to be a partition of the sample set S if $E_1 \cup E_2 \cup \cdots \cup E_n = S$. Use a Venn diagram to show that if $E_1, E_2, \ldots, E_n$ form a partition of S, then any event $A \subset S$ can be split into

$$A = \overset{n}{\underset{i=1}{\uplus}} AE_i$$

where we used $\uplus$ to signify that the union is being formed over a disjoint family of events.

17. Draw a Venn diagram to show that if each occurrence of an event A invariably signifies the occurrence of exactly one event among $E_1, E_2, \ldots, E_n$ so that

$$A = \overset{n}{\underset{i=1}{\uplus}} AE_i$$

this does not necessarily mean that $E_1, E_2, \ldots, E_n$ form a disjoint family, much less that $E_1, E_2, \ldots, E_n$ form a partition.

1.3 THE SIMPLEST PROBABILITY MODEL

In this section we introduce finite probability spaces as the simplest of all mathematical models considered in probability theory. This particular *probability model* draws its basic ingredients from the two preceding sections.

1.3.1 DEFINITION

By a *finite probability space* we mean a triple $(S, \mathscr{S}, P)$ consisting of a finite set S called the *sample set*, a class $\mathscr{S}$ (read script ess) of subsets of S called the *family of considered events*, and the set function P called the *probability measure*, which assigns to each set E belonging to $\mathscr{S}$ a unique real number $P(E)$ called the *probability* of E in such a manner that

(i) $0 \leq P(E)$ and $P(S) = 1$

(ii) $P(A \cup B) = P(A) + P(B)$ if $AB = \phi$ (additivity of P)

The two properties of P above are *axiomatic* in the sense that other properties of P follow from them. We illustrate this by the following example.

Example 1 Derive the following properties of P from properties (i) and (ii) above:

(iii) $P(A^c) = 1 - P(A)$

(iv) $P(A) \leq P(B)$ for $A \subset B$

Solution Since $A^c \cup A = S$, by (i) we have

$$P(A^c \cup A) = 1$$

but since $A^c A = \phi$, by (ii) we obtain

$$P(A^c) + P(A) = 1$$

from which (iii) follows.

As for (iv), since $B = A \cup (B - A)$ where $B - A$ means BA^c, by (ii) we have

$$P(B) = P(A) + P(B - A)$$

but by (i) $P(B - A) \geq 0$, consequently we have

$$P(A) \leq P(B)$$

From a practical point of view the family $\mathscr{S}$ of events in the probability space $(S, \mathscr{S}, P)$ need not include all conceivable events (subsets of S). For as long as $\mathscr{S}$ includes all events of interest to us, $(S, \mathscr{S}, P)$ will serve as a satisfactory probability model. However, it is customary to let $\mathscr{S}$ be inclusive enough so that at least the following conditions are satisfied:

(i)   Whenever two events A and B are in $\mathscr{S}$, so are their union event $A \cup B$ and their intersection event AB.

(ii) Whenever an event A is in $\mathscr{S}$, so is its complement event A^c.

(iii) The sure event S and the impossible event ϕ are in $\mathscr{S}$.

1.3.2 DEFINITION

A family of subsets of S is said to have the structure of a *Boolean algebra* if it is inclusive enough to satisfy the three conditions above.

Stated concisely, our probability space consists of a finite set, a Boolean algebra, and a set function.

Implicit in the specification of a particular probability space is the description of P; a good way of describing P is to indicate $P(E)$ for a smaller collection $\mathscr{S}_0$ of events so that $P(E)$ for every E in $\mathscr{S}$ may be calculated from those $P(E)$ with E in $\mathscr{S}_0$. We might call such a subcollection $\mathscr{S}_0$ a *base* for $\mathscr{S}$ and events in $\mathscr{S}_0$ *basic events*. Consider for instance the following examples.

Example 2 In a thumbtack-tossing experiment if we set $P(H) = 0.4$ we have already completely specified the probability space associated with the experiment since clearly $S = \{h, t\}$, $\mathscr{S} = \{\phi, H, T, S\}$ where $H = \{h\}$, $T = \{t\}$, and $P(\phi) = 0$, $P(H) = 0.4$, $P(T) = 1 - 0.4 = 0.6$, $P(S) = 1$. Thus $\mathscr{S}_0 = \{H\}$ is a base for $\mathscr{S}$.

Example 3 In the die-throwing experiment choices of $p_1, p_2, \ldots, p_6$ for the six outcomes (or elementary events) completely determine the probability space. In general, in any finite probability space the family of elementary events form a base for $\mathscr{S}$. In a special case where all elementary events have equal probabilities, which must then be equal to $1/N$ if N is the number of elements in $\mathscr{S}$, the probability space is completely determined by the number N of outcomes. As we shall see in the next section, such probability models have been extensively used by earlier probabilists.

Example 4 In the experiment of tossing a coin n times, if we let H_i be the event that the ith toss is a head, then $H_1, H_2, \ldots, H_n$ suffice to describe all elementary events; for example we have the elementary event of "all tails" expressed as follows:

$$\{(t, t, \ldots, t)\} = H_1^c H_2^c \cdots H_n^c$$

As we shall see in Chapter 2, the probabilities of these elementary events may be calculated ultimately from the probabilities $P(H_i)$. In this way we see that the events H_i form a base for $\mathscr{S}$. Further if we set $P(H_1) = P(H_2) = \cdots = P(H_n) = p$, say, then we see that a choice of the value p will completely specify the probability space.

It is one thing to specify a probability space $(S, \mathscr{S}, P)$, but quite another to actually determine $P(E)$ for a specific E in $\mathscr{S}$. The adoption of a certain $(S, \mathscr{S}, P)$ amounts to defining the probability of each event represented by some E in $\mathscr{S}$ as $P(E)$. If an event is not represented by any member of $\mathscr{S}$, then we have failed to define the probability for such an event. But even when the probability of an event is defined within the framework of a probability space, it is not in general a trivial matter to actually calculate this probability. For instance, in Example 4 above, is the probability of the event that the coin falls heads exactly r times before it falls tails exactly s times clearly defined, and if so how do we actually calculate it? We shall be able to answer questions like these later.

1.3 PROBLEMS

1. In the experiment of repeatedly tossing a coin n times, if either head or tail repeats itself uninterruptedly for a maximum of $k \leqq n$ times, the event of *a run of size k* is said to occur. Denote such an event by R_k and express R_3 in terms of the basic events $H_1, H_2, \ldots, H_n$ where H_i denotes the event that the ith toss is a head, and $n = 5$.

2. Two unrelated experiments may be combined and considered as a single experiment. If one person tosses a coin and another throws a die, a bystander may observe both experiments and record outcomes, such as $(h, 1)$ or $(t, 6)$. On the other hand, a single experiment may be split and considered as two separate experiments that happen to be performed together. If an experimenter tosses a coin and a thumbtack simultaneously, a bystander may merely observe the coin while another bystander observes the thumbtack. Given two dice to be tossed simultaneously, describe the probability space for each die by listing the six basic (elementary in this case) events, then describe the probability space for both dice by listing twelve basic (but not elementary) events and show how each elementary event can be expressed in terms of these basic events.

3. To an experiment one may associate several probability spaces. Two hardly distinguishable coins are tossed. A gambler observes whether the two faces are alike; a boy observes whether the outcome is two heads, two tails, or one of each; and a teacher observes what happens to each coin. Write out the three probability spaces, and note that an outcome in the gambler's sample set can be traced back to an event in the boy's sample set and that likewise an outcome in the boy's sample set can be traced back to an event in the teacher's sample set. To the four elementary events in the teacher's probability space assign probability values p_1, p_2, p_3, p_4 and determine the probability of every event in the boy's and the gambler's probability spaces.

4. A newly wedded couple plan to have a family of two children. What is the probability of their having one boy and one girl? What sample set did you use? Did you make any unreasonable assumptions in calculating the probability? If three children are planned, what is the probability that all three will be boys?

5. In examining the two-letter words of a certain primitive language, it is found that (1) none of these words consist entirely of consonants, $b, c, d, \ldots$; (2) half of these words begin with a vowel, a, e, i, o, u; and (3) of the words beginning with a vowel only 35% end with a vowel. Consider an appropriate sample set and determine the probability that an arbitrarily picked two-letter word ends with a consonant.

6. In examining the three-letter words of a foreign language, it is found that (1) none of these words consist entirely of vowels; (2) 60% of these words begin with a consonant; (3) if the first letter is a consonant, the odds are 4 to 1 in favor of a vowel for the second letter; and (4) if the first letter is a vowel, the odds are 3 to 1 against a vowel for the second letter; likewise (5) if the second letter is a vowel, the odds are 3 to 1 against a vowel for the third letter. Determine the probability that a three-letter word consists of a consonant, a vowel, and a vowel, in that order.

1.4 THE CLASSICAL FORMULAS

It is sometimes reasonable to make an assumption on statistical regularity without actually performing the random experiment. If a die is not loaded, it is customary to assume that the six outcomes are equally likely. If a coin is not crooked, it is commonly assumed that heads and tails are equally likely. The assumption that all outcomes in a sample set are equally likely may be called the *classical probability assumption*. Historically such an assumption was made untold number of times whenever the context of the experiment warranted it. When a certain 17th-century Frenchman asked Pascal (1623–1662) for advice on the likelihood of at least one "double-6" occurring in 24 throws of a pair of dice, Pascal simply considered all possible outcomes, assumed that they were equally likely, and arrived at the conclusion that the probability was slightly less than half (0.491). We shall find it convenient to call a finite probability space in which all the elementary events have equal probabilities a *classical probability space*.

In a classical probability space of N outcomes each elementary event has the probability equal to $1/N$; if an event E consists of n outcomes, E can be regarded as the disjoint union of n elementary events, hence by the additivity of the probability measure P, we have

$$P(E) = \frac{1}{N} + \frac{1}{N} + \cdots + \frac{1}{N} = \frac{n}{N}$$

We formalize this by stating the following theorem.

1.4.1 THEOREM (The Classical Probability Formula)

If $(S, \mathscr{S}, P)$ is a classical probability space with S consisting of N outcomes, and E is an event consisting of n outcomes, then

$$P(E) = \frac{n}{N}$$

A successful application of the classical probability formula requires that we be able to count outcomes in a given event. We therefore turn to some basic rules and formulas for enumerating the number of elements in a given set.

Addition Rule

The number of elements in the union

$$A_1 \cup A_2 \cup \cdots \cup A_k = \{a : a \in A_i \quad \text{for some} \quad i = 1, 2, \ldots, k\}$$

of pairwise mutually exclusive finite sets $A_1, A_2, \ldots, A_k$ is equal to the sum of the numbers of elements in these sets, that is, if $S = A_1 \cup A_2 \cup \cdots \cup A_k$, then

$$n(S) = n(A_1) + n(A_2) + \cdots + n(A_k)$$

Multiplication Rule

The number of elements in the *Cartesian product*

$$A_1 \times A_2 \times \cdots \times A_k = \{(a_1, a_2, \cdots, a_k) : a_i \in A_i \text{ for each } i = 1, 2, 3 \ldots, k\}$$

of any finite sets $A_1, A_2, \ldots, A_k$ is equal to the product of the numbers of elements in these sets, that is, if $S = A_1 \times A_2 \times \cdots \times A_k$, then

$$n(S) = n(A_1) \times n(A_2) \times \cdots \times n(A_k)$$

Taken by themselves the addition and the multiplication rules are quite trivial. Less trivial is how to recognize situations where these rules may be applied. We illustrate this by the following examples.

Example 1 How many subsets are there in a set of k elements?

Solution Let $\mathscr{S}$ be the family of all subsets of S. Let A_i be the set consisting of two sentences

$$A_i = \{\text{Yes, the } i\text{th element of } S \text{ is in; No, the } i\text{th element of } S \text{ is not in}\}$$

Then $\mathscr{S}$ can be represented as the Cartesian product $A_1 \times A_2 \times \cdots \times A_k$ since each subset of S can be uniquely characterized by an element in the Cartesian product, hence by the multiplication rule we have

$$n(\mathscr{S}) = n(A_1) \cdot n(A_2) \cdot \cdots \cdot n(A_k) = 2^k$$

Example 2 (De Méré's Problem) In an experiment consisting of 24 throws of a pair of dice, how many outcomes belong to the event of at least one double-6.

Solution Let S be the sample set, A the event of at least one double-6, and B the event of no double-6 at all. Then $S = A \cup B$, and by the addition rule we have

$$n(A) = n(S) - n(B)$$

Now, by the multiplication rule

$$n(S) = 36^{24} \qquad n(B) = 35^{24}$$

Hence

$$n(A) = 36^{24} - 35^{24}$$

Incidentally, by using the classical probability formula we can obtain Pascal's result

$$P(A) = \frac{36^{24} - 35^{24}}{36^{24}} = 1 - \left(\frac{35}{36}\right)^{24} \doteq 0.491$$

Permutation Formula

In how many ways can we form a line of r people selected from a group of n people? In other words, how many permutations of size r are possible given n objects to choose from? More briefly how do we calculate P_r^n? It is not difficult to see that

$$P_r^n = n(n-1)(n-2) \cdots (n-r+1) = \frac{n!}{(n-r)!}$$

In particular,

$$P_n^n = n(n-1)(n-2) \cdots 3 \cdot 2 \cdot 1 = n!$$

Example 3 (Abstract Permutations) In lining up all of his n_1 gold coins, n_2 silver coins, and n_3 copper coins, a child wants to know how many distinct "lines" are possible. Naturally the child does not distinguish between coins made of the same metal. He also wants to know if these distinct lines are equally likely when formed at random.

Solution From an adult's point of view each coin is distinct from any other. Hence there are $(n_1 + n_2 + n_3)!$ ways of permuting coins. Each line observed by the child is actually an event consisting of $n_1! \times n_2! \times n_3!$ different outcomes (permutations) since we can permute the coins within each specie without disturbing the line. Hence, dividing the total number of outcomes by the number of outcomes in each event, we obtain the number of lines as

$$\frac{(n_1 + n_2 + n_3)!}{n_1! n_2! n_3!}$$

Clearly all lines are equally likely since they consist of equal number of outcomes that are equally likely.

Example 4 What is the probability that in a given deal each of the four players in a bridge game will be dealt a complete suit of cards?

Solution Suppose we ask each of the four players to return the cards in the order in which they were dealt out to him. If each player was dealt a complete suit, the first 13 cards (coming from the first player) in the deck of 52 cards will be of one suit, and likewise for the next 13 cards, and so on. In general, however, the deck of cards will merely be a permutation of these 52 cards. Since there are 52! such permutations, our sample set S consists of 52! outcomes. Now the event A in which all four players were dealt complete suits can be partitioned into many subevents, such as SHDC and SHCD, where SHDC represents the subevent in which the first player is dealt a complete suit of spades, the second player a complete suit of hearts, and so on. Since there are 4! (permutations of 4 letters) such subevents, and each subevent clearly consists of $13! \times 13! \times 13! \times 13!$ outcomes, the event A consists of $4!(13!)^4$ outcomes. Therefore by the classical probability formula we have

$$P(A) = \frac{4!(13!)^4}{52!} \doteq 0.5 \times 10^{-30}$$

Combination Formula

In how many ways can we form a team of r people selected from a group of n people? In other words, how many combinations of size r are possible given n objects to choose from? More briefly, how do we calculate C_r^n? The formula is derived as follows:

$$C_r^n \times r! = P_s^n \qquad \text{(Why?)}$$

Therefore,

$$C_r^n = \frac{n(n-1)\cdots(n-r+1)}{1 \cdot 2 \cdots \cdot r} = \frac{n!}{(n-r)!\,r!}$$

The numbers C_r^n are referred to as *binomial coefficients* since the multiplication of the binomial $a + b$ by itself n times gives us

$$(a+b)^n = C_n^n a^n + C_{n-1}^n a^{n-1}b + \cdots + C_r^n a^r b^{n-r} + \cdots + C_0^n b^n$$

where the term $a^r b^{n-r}$ is preceded by the coefficient C_r^n denoting the number of different ways in which $a^r b^{n-r}$ emerges in the course of multiplication.

Example 5 (Pascal's Triangle Formula) Prove without calculation the formula

$$C_r^n = C_{r-1}^{n-1} + C_r^{n-1}$$

Solution Each team of r people selected from a group of n people either includes or does not include one particular person (say, yourself). There are exactly C_r^{n-1} teams that do not include that particular person, while there are exactly C_{r-1}^{n-1} teams that include that particular person. By the addition rule the formula is proved.

Example 6 (Constrained Combinations) A group of n children consists of n_1 boys and n_2 girls. In how many ways can we form a team of r children consisting of r_1 boys and r_2 girls?

Solution There are $C_{r_1}^{n_1}$ ways of selecting r_1 boys and $C_{r_2}^{n_2}$ ways of selecting r_2 girls, hence by the multiplication rule there are $C_{r_1}^{n_1} \times C_{r_2}^{n_2}$ ways of forming an acceptable team.

Example 7 What is the probability of your being dealt a complete suit in a bridge game?

Solution As far as you are concerned there are C_{13}^{52} ways of picking out 13 cards to form your hand, of which only 4 ways are acceptable. Consequently the probability is

$$\frac{4}{C_{13}^{52}} = \frac{4(13!)(39!)}{52!} = \frac{(12!)(39!)}{51!}$$

This number is slightly less than 0.6×10^{-11}. Note that the sample set used in this example is less elementary than the one used in Example 4 of this section.

1.4 PROBLEMS

1. Two ordinary dice are rolled. Use the classical probability formula to determine the probability of having a total of 6.

2. Three boys and two girls enter a theater and take five seats at random in a row. What is the probability of the two girls taking the two end seats?

3. A thick coin is tossed 10 times. It falls heads 5 times, tails 3 times, and "sides" 2 times. In how many ways can this occur?

4. Three ordinary dice are thrown and the highest number of dots is observed. If E_i denotes the event that the highest number is i, $n(E_i)$ the number of outcomes in E_i, and $P(E_i)$ the probability of E_i based on the classical probability formula, show that

$$n(E_1) < n(E_2) < n(E_3) < \cdots < n(E_6)$$

to deduce

$$P(E_1) < P(E_2) < \cdots < P(E_6)$$

then actually determine the six probabilities.

5. A fair coin is tossed 10 times. Let E_i denote the event that it falls heads on the ith toss. Are E_1 and its complement E_1^c equally likely? Are $E_1^c E_2$ and $E_1^c E_2^c$ equally likely? What about $E_1^c E_2^c E_3$ and $E_1^c E_2^c E_3^c$? What probabilities would you assign to E_1, $E_1^c E_2$, $E_1^c E_2^c E_3$, and so on? Can you interpret these events verbally?

6. Two pencils are picked out at random from a lot of 7 black and 3 red pencils. What is the probability of an even selection (1 black and 1 red)? If the lot is doubled to 14 black and 6 red pencils, does this change the probability of an even selection? If we keep multiplying the lot by tripling, quadrupling, and so on, does the probability of an even selection approach some specific value?

7. A group of N children includes K boys. A team of $n \leq N$ children is formed by picking from the group at random. Show that the probability of the team containing exactly k boys is given by

$$p_k = \frac{C_k^K C_{n-k}^{N-K}}{C_n^N} \qquad \text{for} \quad k = 0, 1, 2, \ldots, n$$

The probabilities p_0, p_1, p_2, $\ldots$, p_n described above constitute what is known as the *hypergeometric distribution*.

8. A minimum number of gold coins and silver coins are to be put in a pocket so that when two coins are taken out at random the probability that both are gold will be $\frac{1}{2}$. How many of each will be needed? What other mix of coins will give the event the same probability?

9. Prove the following formulas without calculation.
   a. $C_r^n = V_{n-r}^n$.
   b. $C_0^n C_t^n + C_1^n C_{t-1}^n + \cdots + C_t^n C_0^n = C_t^{2n}$, where $0 \leq t \leq n$.

10. Let $a = b = 1$ in the binomial expansion of $(a + b)^n$ to obtain

$$2^n = \sum_{r=0}^{n} C_r^n$$

Regarding 2^n as the number of subsets in a set of n elements, can you interpret the equation as an instance of the addition rule?

1.5 GENERAL PROBABILITY SPACES

For determination of probabilities of events associated with a random experiment having infinitely many outcomes the mathematical model of a finite probability space is no longer adequate. The finiteness condition of the sample set S must be abandoned, the family $\mathscr{S}$ of events must have an additional structure that whenever $A_1, A_2, A_3, \ldots$ are in $\mathscr{S}$ so is their union $\cup_i A_i$, in short $\mathscr{S}$ must be a *sigma* (σ) *algebra* (see Problem 1.5.12 for a formal definition), and finally the probability measure P must have a more extensive additivity property. We shall now give a formal definition of a (general) probability space and explain what this definition entials.

1.5.1 DEFINITION

By a *probability space* we mean a triple $(S, \mathscr{S}, P)$ in which S is a set, $\mathscr{S}$ is a sigma algebra of subsets of S, and P is a set function assigning to each member A of $\mathscr{S}$ a real number $P(A)$ in a manner that ensures:

(i) $0 \leq P(A)$ and $P(S) = 1$
(ii) complete additivity of P: $P(\bigcup_{i=1}^{\infty} A_i) = \sum_{i=1}^{\infty} P(A_i)$ where the sequence of
events A_i form a disjoint family.

Naturally, such a probability space may not come in ready-made for every random phenomenon we encounter. However, acceptance of probability space as the type of mathematical model for investigation of the phenomenon at hand does carry with it a number of basic agreements: for instance, if the probabilities of a certain sequence of disjoint events $A_1, A_2, A_3, \ldots$ are agreed upon, the probability of their union event $A_1 \cup A_2 \cup A_3 \cup \cdots$ will not be arbitrary, but rather be exactly the sum

$$P(A_1) + P(A_2) + P(A_3) + \cdots$$

and this is what condition (ii) means. Two basic agreements that we must accept as consequences of (ii) are stated in the following propositions.

1.5.1 PROPOSITION
If $A_1 \subset A_2 \subset A_3 \subset \cdots$ form an increasing sequence of events, and if

$$A = \bigcup_{i=1}^{\infty} A_i$$

then

$$P(A) = \lim_{i \to \infty} P(A_i)$$

In other words, $P(A)$ can not be arbitrary relative to $P(A_i)$.

PROOF
Let $D_i = A_i - A_{i-1}$, then clearly (draw a Venn diagram)

$$A = \bigcup_{i=1}^{\infty} D_i$$

Hence, by the complete additivity of P we must have

$$P(A) = \sum_{i=1}^{\infty} P(D_i)$$

$$= \lim_{n \to \infty} \sum_{i=1}^{n} P(D_i)$$

but

$$\sum_{i=1}^{n} P(D_i) = P(D_1 \cup D_2 \cup \cdots \cup D_n) = P(A_n)$$

and so

$$P(A) = \lim_{n \to \infty} P(A_n)$$

1.5.2 PROPOSITION

If $A_1 \supset A_2 \supset A_3 \supset \cdots$ form a decreasing sequence of events, and if

$$A = \bigcap_{i=1}^{\infty} A_i$$

then

$$P(A) = \lim_{i \to \infty} P(A_i)$$

Note that the intersection of a sequence of decreasing sets can be either empty or nonempty. Consider for example the sequence of intervals $A_i = [i, \infty)$ and also the sequence $B_i = [0, 1 + 1/i]$.

PROOF

Let $D_i = A_i - A_{i+1}$, then clearly (again draw a Venn diagram)

$$A \cup D_1 \cup D_2 \cup \cdots = A_1$$

and again by the complete additivity of P we have

$$P(A) + \sum_{i=1}^{\infty} P(D_i) = P(A_1)$$

or

$$P(A) = P(A_1) - \lim_{n \to \infty} \sum_{i=1}^{n} P(D_i)$$

$$= \lim_{n \to \infty} \left[P(A_1) - \sum_{i=1}^{n} P(D_i) \right]$$

but

$$P(A_1) - \sum_{i=1}^{n} P(D_i) = P(A_{n+1})$$

from which we have

$$P(A) = \lim_{n \to \infty} P(A_n)$$

In determining probabilities of events associated with a given random experiment we proceed roughly as follows: First, a sample set S is specified, subsets of which are used to represent various events. Next, a number of basic events whose probabilities can be readily agreed upon within the context of the experiment are actually represented by specific subsets of S. Finally, all the events whose probabilities cannot be arbitrary relative to the probabilities of these basic events are represented and grouped together to form the family (of subsets or events) $\mathscr{S}$, which can be shown to have the structure of a sigma algebra. If events of interest to us belong to $\mathscr{S}$, then we have chosen our basic events well, for the probabilities of these less basic events are now well defined and it remains to actually calculate them.

Although we associate with each random experiment a probability space, in practice we need not bother with detailed construction of such a model. Quite often we regard a probability space merely as a theoretical reference frame within which to calculate probabilities of events that are of interest to us. In other words, subsequent to a recognition of the sample set and an agreement on the probabilities of a few basic events we may proceed directly to the calculation of the probability of a particular event.

Example 1 A coin with an extremely small probability, say, $p = 0.5 \times 10^{-30}$, of falling heads is tossed repeatedly till a head appears. Within the framework of a suitable probability space determine the probability of the event that the tossing terminates sooner or later.

Solution First we consider the infinite sample set

$$S = \{h, th, tth, ttth, \ldots\}$$

and let the event that the tossing terminates at the first toss be represented by $A_1 = \{h\}$, that the tossing terminates at the second toss by $A_2 = \{th\}$, and so on. Then the event we are interested in can be represented by

$$A = A_1 \cup A_2 \cup A_3 \cup \cdots$$

Since the events A_i clearly form a disjoint family, by the complete additivity of P we have

$$P(A) = P(A_1) + P(A_2) + \cdots$$

Thus if the basic probabilities $P(A_1)$, $P(A_2)$, ... are chosen, $P(A)$ should be just the sum of these probabilities. One set of reasonable choices would be $P(A_1) = p$, $P(A_2) = qp, \ldots, P(A_n) = q^{n-1}p, \ldots$, where $q = 1 - p$, so that

$$P(A) = p + qp + q^2p + \cdots$$

$$= p \sum_{n=0}^{\infty} q^n$$

$$= p \frac{1}{1-q} = 1$$

A quicker way to arrive at $P(A) = 1$ is to observe that

$$A = A_1 \cup A_2 \cup \cdots = S$$

and

$$P(S) = 1$$

by (i) of Definition 1.5.1.

Note that the probabilities $P(A_1)$, $P(A_2)$, ... above can be tested empirically while $P(A)$ cannot be since the time required for one single performance of the experiment is not bounded. However, the acceptance of the particular probability space as the mathematical model dictates that the probability of the event in question is exactly equal to 1 (whether we like it or not). If a different probability space is considered, the probability of this same event may not be so obviously equal to 1. Consider for example as the sample set of the same experiment the set S of all infinite sequences in two letters h and t, and represent the event that the tossing terminates at the nth toss by the subset A_n of S consisting of all the sequences in which the first $n - 1$ letters are t and the nth letter is h. Again A_n will form a disjoint family of subsets of S, and if we let $A = \bigcup_n A_n$ represent the event that the tossing terminates sooner or later, then A is no longer the sure event S as it was in the previous sample set since now the sequence $(t, t, t, \ldots)$ in which h never appears does not belong to any A_n and hence not to A. However, if we assign the probability $q^{n-1}p$ to A_n, which is the only reasonable thing to do from an empirical point of view, then again the probability of A must be equal to 1.

The two probability spaces we have just considered for the experiment of tossing a coin till a head appears are of fundamentally different nature since the first probability space has a sample set which is *countable* in the sense that all the elements in it can be lined up exhaustibly in a sequence while the second probability space has a sample set which is *noncountable* in the sense that every sequence of elements is bound to miss out some particular elements (see Problem 1.5.2).

A probability space in which the sample set is countable is called a *discrete probability space*. A discrete probability space is completely described by specifying probability values for all the elementary events (or the individual outcomes if one likes to think of it that way). Incidentally, a finite probability space is a particular case of discrete probability space since a finite set is regarded as countable. Although discrete probability spaces constitute only an elementary subcollection of all probability spaces, surprisingly many random phenomena can be studied under this type of probability spaces (see Minibibliography, Feller, vol. 1).

A probability space $(S, \mathscr{S}, P)$ in which S is noncountable may still be quite elementary if $\mathscr{S}$ is kept simple. Thus what makes a probability space complex is not so much the enormity of S as of $\mathscr{S}$. The following problem can be worked out by using a fairly elementary nondiscrete probability space.

Example 2 (Mosteller) On a large checkerboard a coin of diameter 0.5 in. is tossed. How small must the squares be in order that the probability of the coin landing clear of all lines be as small as 0.01?

Solution If the sides of the squares are too small, say, $x \leqq 0.5$, then the probability of the coin landing clear of the lines is 0. Therefore what we should do is to make x slightly larger than $d = 0.5$.

We choose as our sample set any one of the squares since all squares are alike and each point in the square is a potential landing place for the center of the coin. Now in order that the coin lands clear of the sides of the square the center of the coin must end up in the inner portion A of the square (see Fig. 1.5.1). Since the coin is "uniformly likely" to land anywhere in the square, it is reasonable to let

$$P(A) = \frac{\alpha(A)}{\alpha(S)}$$

where $\alpha(A)$ and $\alpha(S)$ are respectively the areas of A and S. Such a choice of value of $P(A)$ also commit us to a specific value for $P(S - A)$, the probability of the event that the coin rests on the lines since

$$P(S - A) = 1 - P(A) = 1 - \frac{\alpha(A)}{\alpha(S)}$$

$$= \frac{\alpha(S) - \alpha(A)}{\alpha(S)}$$

$$= \frac{\alpha(S - A)}{\alpha(S)}$$

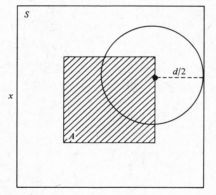

Figure 1.5.1

Returning to $P(A)$, we set

$$P(A) = \frac{(x - d)^2}{x^2} = 0.01$$

Letting $d = 0.5$, we solve for x to obtain

$$x = \tfrac{5}{9} \doteq 0.555 \cdots$$

We point out the formal similarity between $P(A) = \alpha(A)/\alpha(S)$ and the earlier $P(A) = n(A)/n(S)$. Note that counting of the number of elements in A has been replaced by measuring of the area of A, and the phrase "equally likely" has been replaced by "uniformly likely." The solution of the following well-known problem is yet another application of the formula $P(A) = \alpha(A)/\alpha(S)$.

Example 3 (Buffon's Needle Problem) A needle of L in. is dropped at random on a large piece of paper ruled with lines that are $D > L$ in. apart. Determine the probability of the needle touching one of the lines.

Solution Clearly the needle can touch at most one line since $L < D$, and whether the needle touches one of the lines is completely determined by (1) the distance x (inches) from the midpoint of the needle to the nearest line and (2) the acute angle θ (radians) formed by the needle and the line (see Fig. 1.5.2a).

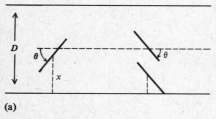 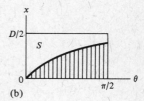

(a)   (b)

Figure 1.5.2

We take as our sample set the rectangle (see Fig. 1.5.2b)

$$S = [0, \pi/2] \times [0, D/2]$$

since θ ranges in the interval $[0, \pi/2]$ while x ranges in the interval $[0, D/2]$. Now for each fixed θ, x has to be sufficiently small in order for the needle to touch the line (see Fig. 1.5.2a). In fact we must have

$$x \leq \frac{L}{2} \sin \theta$$

Consequently the event that the needle touches a line must be represented by the portion A of S under the sine curve $x = (L/2) \sin \theta$ (see Fig. 1.5.2b). Now the areas $\alpha(S)$ and $\alpha(A)$ are $\pi D/4$ and $L/2$, respectively, as the reader can easily check, if he is so inclined, with a bit of integral calculus. Therefore,

$$P(A) = \frac{\alpha(A)}{\alpha(S)} = \frac{2L}{\pi D}$$

We point out that the values of x in the interval $[0, D/2]$ are "uniformly likely" and, for each fixed value of x, the values of θ in the interval $[0, \pi/2]$ are again "uniformly likely" so that all points in S are "uniformly likely," and this "justifies" our using the area-ratio formula above. Finally we remark that for $L = D$ we have $P(A) = 2/\pi$. This last result enables us to estimate the value of π in terms of the frequency ratio of the needle touching the lines in a long series of the experiment.

1.5 PROBLEMS

1. Show that the set of all rational numbers is countable by first listing the positive ones as follows and counting them off in the order indicated by the arrows:

$$1/1 \rightarrow 1/2 \quad 1/3 \rightarrow 1/4 \quad 1/5 \cdots$$
$$2/1 \quad 2/2 \quad 2/3 \quad 2/4 \quad 2/5 \cdots$$
$$3/1 \quad 3/2 \quad 3/3 \quad 3/4 \quad 3/5 \cdots$$
$$4/1 \quad 4/2 \quad 4/3 \quad 4/4 \quad 4/5 \cdots$$

2. Given a sequence of sequences in two letters h and t, construct a sequence in h and t by choosing the first letter to be different from the first letter of the first sequence, the second letter to be different from the second letter of the second sequence, and so on. Show that the sequence so constructed is not one of the sequences in the given sequence of sequences. Prove that the set of all real numbers in the unit interval [0, 1] is not countable. (Hint: Represent each real number by its decimal expansion.)

3. Let a certain random experiment be described by $(S, \mathscr{S}, P)$ where S is the set of all positive rational numbers

$$S = \{r_1, r_2, r_3, \ldots\}$$

$\mathscr{S}$ is the family of all subsets of S, and P is such that for any $A \subset S$

$$P(A) = \sum_i p_i \quad \text{(summation over all } i \text{ such that } r_i \in A)$$

where $\{p_1, p_2, p_3, \ldots\}$ is a sequence of nonnegative real numbers such that their sum is equal to 1. Is $(S, \mathscr{S}, P)$ a probability space in the sense of Definition 1.5.1? Is it a discrete probability space?

4. A 1-ft tape is painted alternatively red and green along the decreasing portions of lengths $\frac{1}{2}$ ft, $\frac{1}{4}$ ft, $\frac{1}{8}$ ft, $\frac{1}{16}$ ft, ... A dart is thrown randomly at the tape. Determine the probability that it hits the red portion of the tape.

5. A coin of diameter 1 in. is tossed at random onto a large checkerboard with lines 3 in. apart. Determine the probability that the coin lands clear of all intersections (corners).

6. A dart is thrown at random at a square target. Use Proposition 1.5.2 to show that the probability of the dart hitting the exact center is 0.

7. Two friends agree to take a chance at meeting at the airport between 12 noon and 1 P.M. Each agrees to come to the airport during the hour and wait for 20 minutes. What is the probability of their actually meeting?

8. A stick is broken into three pieces at random. What is the probability that the three pieces will form a triangle? Hint: Each piece has to be less than half the length of the stick.

9. Let $A_1, A_2, A_3, \ldots$ be a sequence of events. Prove the following so-called Boole's inequality:

$$P\left(\bigcup_i A_i\right) \leq \sum_i P(A_i)$$

Hint: Let $D_i = A_i - (A_1 \cup A_2 \cup \cdots \cup A_{i-1})$, then $\bigcup_i A_i = \bigcup_i D_i$.

10. If $A_1 \supset A_2 \supset A_3 \supset \cdots$ are such that $\bigcap_i A_i = \phi$, show that

$$\lim_{i \to \infty} P(A_i) = 0$$

11. Use De Morgan's formulas:

$$\left(\bigcup_i A_i\right)' = \bigcap_i A_i'$$

$$\left(\bigcap_i A_i\right)' = \bigcup_i A_i'$$

to show that if $A_1, A_2, A_3, \ldots$ with $\bigcap_i A_i = A$ is a decreasing sequence of events, then $A_1', A_3', A_2', \ldots$ is an increasing sequence of events with $\bigcup_i A_i' = A'$. Derive Proposition 1.5.2 from Proposition 1.5.1.

12. The formal definition of a sigma algebra is as follows. A nonempty family $\mathscr{S}$ of subsets of a given set S is said to be a *sigma algebra* if it satisfies the following two conditions:

   i. Whenever a subset A is a member of $\mathscr{S}$, so is its complement A'.

   ii. Whenever the subsets $A_1, A_2, \ldots$ forming a sequence (finite or infinite) are all members of $\mathscr{S}$, so is their union $\bigcup_i A_i$.

Prove that (i) and (ii) above imply the following:

   iii. S and ϕ are both members of $\mathscr{S}$.

   iv. If the subsets $A_1, A_2, \ldots$ forming a sequence are all members of $\mathscr{S}$, so is their intersection $\bigcap_i A_i$. Hint: Use De Morgan's formula.

   v. If the subsets $A_1, A_2, \ldots, A_n$ forming a finite sequence are all members of $\mathscr{S}$, so are their union and their intersection. Hint: Let $A_{n+1} = A_{n+2} = \cdots = \phi$.

13. Show that a finite probability space (Definition 1.3.1) is a probability space in the sense of Definition 1.5.1 by showing that a finite Boolean algebra is automatically a sigma algebra, and that for a finite Boolean algebra the complete additivity of P is trivially satisfied.

14. In the experiment of breaking a teapot to pieces let A_i be the event that the ith heaviest piece weighs less than some fixed $\epsilon < 0$. Show that $A_1 \subset A_2 \subset A_3 \cdots$ and that $\bigcup_i A_i = S$ where S is the sure event.

Chapter 2
CONDITIONAL PROBABILITY AND SPECIAL PROBABILITY SPACES

Probability spaces are not created just for aesthetic reasons; they have to enable us to determine probabilities of random events. In constructing a probability space, the determination of the probability measure is always the most important, and often the most difficult. The concept of conditional probability introduced in Section 2.1 (The Idea of Conditional Probability) plays an essential part in overcoming this difficulty; its effectiveness is summarized in several useful formulas appearing in Section 2.2 (Conditional Probability Formulas).

After considering a very important special situation in Section 2.3 (Stochastic Independence of Events), which arises from the general consideration of conditional probability, the newly developed tools are used in constructing three special classes of probability spaces. These appear in Sections 2.4–2.6 (Bernoulli, Markov, and Poisson Probability Spaces). Part or all of these three sections may be postponed since they are not used later in the book. If these sections are to be omitted entirely, the exercise problems of Section 2.3 about the finite Bernoulli probability spaces should be given sufficient attention.

Section 2.3 is by far the most important in the chapter; the concept of stochastic independence will be repeatedly used to describe relatively uncomplicated situations for which many relatively simple mathematical results are known.

2.1 THE IDEA OF CONDITIONAL PROBABILITY

Given a probability space $(S, \mathscr{S}, P)$ for some random phenomenon, a change in observer's point of view may lead to an alteration in S and the corresponding modifications in $\mathscr{S}$ and P. If for some reason certain outcomes in the original sample set are deliberately excluded from consideration so that the remaining outcomes constitute a subset A of S, then an event formerly represented by a subset B should now be represented by BA (read B intersection A) and the probability of this event, which was formerly $P(B)$, should now perhaps be modified to $P(BA)$; only such a modification is not completely satisfactory since $P(BA) \leq P(A)$, which means that no event, not even the *new* sure event A, will have the probability value 1 unless $P(A)$ happens to be 1, which is in general not the case. This leads us to consider scaling up the probability value $P(BA)$ by the magnifying factor $1/P(A) \geq 1$, that is we shall let $P(BA)/P(A)$ be the new revised probability value and call it the conditional probability of the event B given the event A. We now state formally the following definition.

2.1.1 DEFINITION

Given a probability space $(S, \mathscr{S}, P)$ and a fixed event $A \in \mathscr{S}$, we define for any event $B \in \mathscr{S}$ the *conditional probability of B given A* by

$$P(B|A) = \frac{P(BA)}{P(A)}$$

Obviously we must require that $P(A) > 0$. That is, $P(B|A)$ is not defined unless $P(A) > 0$. Note also if $A = S$, we have $P(B|S) = P(B)$.

Example 1 Two ordinary dice are tossed behind you; you are then informed that the total number of dots is not greater than three. Find first the probability of the event of "two faces alike" ignoring the information given, then revise your probability in view of the information given.

Solution If $B = \{(1, 1), (2, 2), \ldots, (6, 6)\}$, then clearly $P(B) = \frac{6}{36} = \frac{1}{6}$. Now, in view of the information given your sample set narrows down to $A = \{(1, 1), (1, 2), (2, 1)\}$ consisting of 3 outcomes instead of the original 36. The event of "two faces alike" also narrows down to $\{(1, 1)\} = BA$. Hence

$$P(B|A) = \frac{P(BA)}{P(A)} = \frac{1/36}{3/36} = \frac{1}{3}$$

Thus the "hope" for "two faces alike" has doubled from $\frac{1}{6}$ to $\frac{1}{3}$ in view of the information given.

Needless to say, the probability of a given event may be decreased as well as increased by the given information. For example, if B is the event of an "even dots" and A the event of an "odd dots" in the die throwing experiment, then clearly $P(B|A) = 0$ since one event completely excludes the other.

Example 2 In a game of bridge if you are dealt a complete suit, what is the probability of your partner also being dealt a complete suit?

Solution Since you have been dealt a complete suit, the rest of the cards consist of three suits totaling 39 cards. Now as far as your partner is concerned, there are C_{13}^{39} ways of forming a hand, of which only 3 ways result in a complete suit. Therefore the conditional probability of the event B (your partner being dealt a complete suit) given the event A (your being dealt a complete suit) is

$$P(B|A) = \frac{3}{C_{13}^{39}} = \frac{12!26!}{38!}$$

The preceding example serves to point out that the conditional probability may sometimes be figured out without recourse to the original formula $P(B|A) = P(BA)/P(A)$. Whenever this is the case, and this is precisely what makes the concept of conditional probability such a useful one, the above formula may be rewritten as the *product formula*

$$P(BA) = P(A)P(B|A)$$

to find $P(BA)$ from $P(A)$ and $P(B|A)$. We illustrate this by the following example.

Example 3 Two socks are drawn from a drawer containing 3 black and 2 red socks. Find the probability of drawing a red pair.

Solution Suppose we draw one sock at a time, and let R_1 be the event that the first sock drawn is red and R_2 the event that the second sock drawn is red. By the product formula we have

$$P(R_1 R_2) = P(R_1)P(R_2|R_1)$$

Now $P(R_1) = \frac{2}{5}$ since 2 out of the 5 socks we draw from are red, and $P(R_2|R_1) = \frac{1}{4}$ since after one red sock is drawn only 1 out of the remaining 4 is red. Consequently,

$$P(R_1 R_2) = \frac{2}{5} \times \frac{1}{4} = \frac{1}{10}$$

2.1 PROBLEMS

1. A gambler bet on two heads before tossing two fair coins. After the tossing he was informed that both coins had fallen on the same face. Although it was too early to rejoice since he was yet to be informed of the actual outcome, he figured that his chance of winning the bet was now improved somewhat. Calculate the probabilities of his winning before and after the preliminary information given.

2. A dart is thrown at random on a 4-ft-square board with a circular target of radius 1 ft in the center. What is the probability that it hits the target? What is the probability that it hits the target given that it has hit the upper half of the board; given that it has hit the upper quarter of the board; given that it has hit the middle half of the board; given that it has hit the middle hundredth of the board?

3. Consider a deck of three cards one of which is painted red on both sides, another white on both sides, and the third red on one side and white on the other. One of these cards is picked out at random and laid on the table. If it shows red, what is the probability that the other side is also red? Hint: Work with a specific sample set. Note that $\frac{1}{2}$ is a common wrong answer.

4. A stick is broken into three pieces at random. Let A be the event that the left-end piece is of more than half length, and B be the event that the middle piece is of more than a quarter length. Calculate the probability $P(B)$ and the conditional probability $P(B|A)$. Are these probabilities somewhat as you expected?

5. A problem in conditional probability may be equivalent to another in ordinary probability:
   a. Three dice are rolled. What is the probability that one of the three faces is a 6 given that all three faces are different?
   b. Three cards are drawn simultaneously from a deck of six cards labeled 1, 2, 3, 4, 5, 6. What is the probability that a 6 is among those drawn?

6. Prove that if $P(B|A) > P(B)$, then $P(B^c|A) < P(B^c)$. How would you interpret this result? If $A \subset B$, show that $P(B|A) = 1$. Can you see why the converse is not quite true? If $AB = \phi$, show that $P(B|A) = 0$. What is your interpretation?

7. Three relations are possible between $P(B|A)$ and $P(B)$, namely

$$P(B|A) > P(B) \quad \text{or} \quad P(B|A) = P(B) \quad \text{or} \quad P(B|A) < P(B)$$

How would you interpret these three relations?

8. Although we have defined conditional probability $P(\cdot|A)$ in terms of ordinary probability $P(\cdot)$, we could also regard $P(\cdot|A)$ as an abstraction of some "conditional frequency ratio." The frequency ratio $f_n(B)$ of an event B was defined as the ratio of the number of occurrences of B to the total number n of performances of the experiment. How would you define the *conditional frequency ratio* $f_n(B|A)$ of an event B given another event A? If $A = S$, does $f_n(B|S)$ agree with $f_n(B)$?

9. Given a probability space $(S, \mathcal{S}, P)$ and A with $P(A) > 0$, can you define a new probability space having A as the sample set and the conditional probability $P(\cdot|A)$ as the probability measure? What would your new family of events (denote it by $\mathcal{S}_A$) be like? Can you prove that $[A, \mathcal{S}_A, P(\cdot|A)]$ is indeed a probability space?

10. Given a probability space $(S, \mathcal{S}, P)$ and A with $P(A) > 0$, show that the set function P_A on $\mathcal{S}$ defined by

$$P_A(E) = P(E|A) \qquad \text{for any } E \text{ in } \mathcal{S}$$

satisfies the two axioms of probability measure and that consequently $(S, \mathcal{S}, P_A)$ is a probability space also. (Cf. Definition 1.5.1.)

2.2 CONDITIONAL PROBABILITY FORMULAS

The real usefulness of conditional probability is found, as was pointed out in the preceding section, in the product formula

$$P(AB) = P(A)P(B|A)$$

This basic product formula can be used to derive a number of additional useful formulas (see Propositions 2.2.1–4). The following *fundamental lemma* concerning conditional probability will be helpful in deriving the propositions.

2.2.1 LEMMA

Given a probability space $(S, \mathscr{S}, P)$ and $A \in \mathscr{S}$ with $P(A) > 0$, if we define the set function P_A by

$$P_A(B) = P(B|A) \qquad \text{for any} \quad B \in \mathscr{S}$$

then P_A is a probability measure so that $(S, \mathscr{S}, P_A)$ is a probability space. Furthermore, if $P_A(B) > 0$, then

$$P_A(C|B) = P(C|AB) \qquad \text{for any} \quad C \in \mathscr{S}$$

PROOF

It can be easily shown that P_A satisfies the axioms of probability measure (cf. Definition 1.5.1). Since $(S, \mathscr{S}, P_A)$ is a probability space in its own right, we may consider for any given $B \in \mathscr{S}$ with $P_A(B) > 0$ the conditional probability $P_A(\cdot|B)$ exactly as in Definition 2.1.1. Now to show $P_A(C|B) = P(C|AB)$, we have

$$P_A(C|B) = \frac{P_A(CB)}{P_A(B)} = \frac{P(CB|A)}{P(B|A)} = \frac{P(CBA)/P(A)}{P(BA)/P(A)}$$

$$= \frac{P(CBA)}{P(BA)} = P(C|AB)$$

and this completes the proof.

In view of the preceding lemma we can derive a generalization of the basic product formula, $P(AB) = P(A)P(B|A)$, which after all deals with intersection of only two events. Consider now $P(A_1 A_2 \cdots A_n)$. We see that

$$P(A_1 A_2 \cdots A_n) = P[A_1(A_2 \cdots A_n)]$$
$$= P(A_1)P(A_2 \cdots A_n|A_1)$$

but

$$P(A_2 \cdots A_n|A_1) = P_{A_1}[A_2(A_3 \cdots A_n)]$$
$$= P_{A_1}(A_2)P_{A_1}(A_3 \cdots A_n|A_2)$$
$$= P(A_2|A_1)P(A_3 \cdots A_n|A_1 A_2)$$

so that

$$P(A_1 A_2 \cdots A_n) = P(A_1)P(A_2 | A_1)P(A_3 \cdots A_n | A_1 A_2)$$

Working on the last term or continuing inductively, we arrive at the following proposition.

2.2.1 PROPOSITION (Extended Product Formula)

Given a probability space $(S, \mathscr{S}, P)$, if $A_1, A_2, A_3, \ldots, A_n$ are events such that all the conditional probabilities appearing below exist, then

$$P(A_1 A_2 \cdots A_n) = P(A_1)P(A_2 | A_1)P(A_3 | A_1 A_2) \cdots P(A_n | A_1 A_2 \cdots A_{n-1})$$

Although our derivation preceding the statement of this proposition constitutes a proof of this formula, we give instead the following simpler proof.

PROOF

Looking at the right-hand side of the equation above, we see that

$$P(A_1)P(A_2 | A_1) = P(A_1 A_2)$$

by the basic product formula, but then

$$P(A_1 A_2)P(A_3 | A_1 A_2) = P(A_1 A_2 A_3)$$

again by the basic product formula. Continuing this way we see the entire right-hand side of the equation collapses into $P(A_1 A_2 A_3 \cdots A_n)$.

Let us work out a simple example by using Proposition 2.2.1.

Example 1 Three boys and two girls enter a theater and take five seats at random in a row. What is the probability that the girls take the two end seats?

Solution Labeling the seats by 1, 2, 3, 4, 5, and letting B_i and G_i respectively denote the event that the ith seat is taken by a boy and by a girl, we want to calculate the probability of the event $G_1 B_2 B_3 B_4 G_5$. Now by the extended product formula we have

$$P(G_1 B_2 B_3 B_4 G_5) = P(G_1)P(B_2 | G_1)P(B_3 | G_1 B_2)P(B_4 | G_1 B_2 B_3)P(G_5 | G_1 B_2 B_3 B_4)$$

The five (conditional) probabilities appearing on the right-hand side can be figured out expeditiously as $\frac{2}{5}, \frac{3}{4}, \frac{2}{3}, \frac{1}{2}$, and 1. Therefore,

$$P(G_1 B_2 B_3 B_4 G_5) = \frac{2}{5}, \frac{3}{4}, \frac{2}{3}, \frac{1}{2}, 1 = \frac{1}{10}$$

Incidentally, if we wrote the event in question as $G_1 G_5 B_2 B_3 B_4$ (remembering the commutative law), then we would have equivalently

$$P(G_1 G_5 B_2 B_3 B_4) = \frac{2}{5}, \frac{1}{4}, \frac{3}{3}, \frac{2}{2}, 1 = \frac{1}{10}$$

The next formula, which is often called the "formula for total probability," is the result of applying the basic product formula several times.

2.2.2 PROPOSITION (Partition Formula)

Given a probability space $(S, \mathscr{S}, P)$ and a partition of S by n events, $A_1, A_2, A_3, \ldots, A_n$, we have for any event B

$$P(B) = \sum_{i=1}^{n} P(A_i)P(B|A_i)$$

assuming of course that $P(B|A_i)$ is defined or $P(A_i) > 0$ for each i.

PROOF

From $B \subset \bigcup_i A_i$ follows (draw a Venn diagram if necessary)

$$B = \bigcup_i BA_i$$

By additivity of P we obtain

$$P(B) = \sum_i P(BA_i)$$

Now applying the basic product formula to each $P(BA_i)$ we see

$$P(B) = \sum_i P(A_i)P(B|A_i)$$

Example 2 In a multiple-choice type test each question is followed by one correct and four incorrect answers. A student naturally picks the correct answer when he really knows the answer, but picks an answer at random when he is totally ignorant of the answer. We will assume that he is invariably trapped by a deceptive incorrect answer when he is half-sure of himself and so hazards a guess. The teacher estimates that a certain student really knows the answer for 50% of the questions, and is totally ignorant about 10% of the questions. The student will hazard guesses for the rest of the questions. What is the probability that the student picks the correct answer for, say, question 17?

Solution Confronted with question 17 the student can react in the following six logically possible ways (outcomes). Thus the sample set consists of

(a_1, b):   "knows the answer and picks the correct answer"
(a_1, d):   "knows the answer but picks an incorrect answer"
(a_2, b):   "is totally ignorant but picks the correct answer"
(a_2, d):   "is totally ignorant and picks an incorrect answer"
(a_3, b):   "is half-sure of himself and picks the correct answer"
(a_3, d):   "is half-sure of himself and picks an incorrect answer"

If we let A_1, A_2, A_3 be respectively the events that the student knows the answer, is totally ignorant, and is half-sure of himself, then these events form a partition of the sample set S, each event consisting of exactly two outcomes. The teacher's estimate of the student's ability may be translated into

$$P(A_1) = 0.5$$
$$P(A_2) = 0.1$$
$$P(A_3) = 0.4$$

Now if B is the event that the student picks the correct answer, we can expect

$$P(B|A_1) = 1$$
$$P(B|A_2) = 0.2$$
$$P(B|A_3) = 0$$

Consequently, using the partition formula, we have

$$P(B) = \sum_{i=1}^{3} P(A_i)P(B|A_i) = (0.5)(1) + (0.1)(0.2) + (0.4)(0)$$
$$= 0.52$$

Next we introduce the well-known Bayes formula, which is a simple consequence of combining the partition formula and the earlier product formula. Simply stated, Bayes's formula enables us to use the information $P(A_1)$, $P(A_2)$, ..., $P(A_n)$, and $P(B|A_1)$, $P(B|A_2)$, ..., $P(B|A_n)$ to calculate $P(A_1|B)$, $P(A_2|B)$, ..., $P(A_n|B)$, thus leading us from one set of conditional probabilities to another set of conditional probabilities.

2.2.3 PROPOSITION (Bayes's Formula)
Given a probability space $(S, \mathscr{S}, P)$ and a partition of S by n events $A_1, A_2, \ldots, A_n$, we have for any event B

$$P(A_i|B) = \frac{P(A_i)P(B|A_i)}{\sum_j P(A_j)P(B|A_j)}$$

for $i = 1, 2, \ldots, n$, assuming of course $P(A_j) > 0$ and $P(B) > 0$.

PROOF
Let us consider, say, $P(A_1|B)$. By definition

$$P(A_1|B) = \frac{P(A_1B)}{P(B)}$$

Applying the product formula to the numerator and the partition formula to the denominator, we obtain at once

$$P(A_1|B) = \frac{P(A_1)P(B|A_1)}{\sum_j P(A_j)P(B|A_j)}$$

Example 3 (Continuation of Example 2) If the student has picked the correct answer for question 17, what is the probability that he really knew the answer?

Solution We need only calculate $P(A_1 | B)$ by Bayes's formula. But since Bayes's formula is so easily derived from product formula and partition formula we need not rush to memorize it. Now since

$$P(A_1 | B) = \frac{P(A_1)P(B | A_1)}{P(B)}$$

and $P(A_1) = 0.5$, $P(B | A_1) = 1$, and the total probability $P(B)$ was already calculated to be 0.52, we obtain $P(A_1 | B) = 0.5/0.52 \doteq 0.96$.

Thus it seems the student most likely knew the answer and was not merely guessing. It is not hard to see that as the number of choices tends to infinity, $P(A_1 | B)$ will approach 1 (see Problem 2.2.4).

Bayes's formula is considered useful in calculating the so-called a posteriori probabilities such as $P(A_1 | B)$, $P(A_2 | B)$, $P(A_3 | B)$, ..., into which the original a priori probabilities $P(A_1)$, $P(A_2)$, $P(A_3)$, ... are revised *after* (hence a posteriori) the occurrence of the event B.

We conclude this section with a simple generalization of the partition formula. The resulting formula, which we shall call the *conditional partition formula*, will be used only in the last three sections of this chapter.

2.2.4 PROPOSITION (Conditional Partition Formula)
Given a probability space $(S, \mathscr{S}, P)$ and a partition of S by n events, $A_1, A_2, \ldots,$ A_n, we have for any event B and for any fixed event D

$$P(B | D) = \sum_i P(A_i | D)P(B | A_i D)$$

assuming as usual that all conditional probabilities involved exist.

Note that if in particular $D = S$, the conditional partition formula reduces to the ordinary partition formula.

PROOF
In view of Lemma 2.2.1 and the partition formula we have

$$P(B | D) = P_D(B) = \sum_i P_D(A_i)P_D(B | A_i)$$
$$= \sum_i P(A_i | D)P(B | A_i D)$$

completing the proof.

2.2 PROBLEMS

1. Suppose a jar contains 3 white balls and 1 black ball. Three balls are drawn from the jar and placed on the table in the order drawn. What is the probability that the middle one is black?

2. Three teachers and three pupils take seats around a round table at random. What is the probability that the teachers and the pupils take the alternate seats?

3. A lens manufactured by a certain optical company breaks $\frac{1}{2}$ of the time when dropped the first time, $\frac{3}{10}$ of the time when dropped the second time, and $\frac{9}{10}$ of the time when dropped the third time. If such a lens is dropped three times, what is the probability that it will break?

4. In Section 2.2, Example 3, show that as the number of answers to choose from is increased indefinitely the conditional probability $P(A_1|B)$ approaches 1. Hint: First show that $P(A_2|B)$ approaches 0.

5. Mr. Moody is either moody or cheerful throughout the day. He is always moody on Monday. The probability of his remaining moody the day following a moody day is 0.9, and the probability of his remaining cheerful the day following a cheerful day is 0.3. Calculate the probability of his being moody on Wednesday.

6. Two messages coded as A and B are transmitted to the receiving station where A is received mistakenly as B with probability 0.02 and B is received mistakenly as A with probability 0.01. Message A is sent out twice as frequently as B. Message A is received; what is the probability that it is really message A?

7.   According to a Kaiser Hospital report 45% of the babies delivered last year were boys, 70% of the baby boys were born with black hair, and 50% of the baby girls were born with black hair. A baby born last year is to be adopted by a certain family. Given that the baby was born with black hair, what is the probability that the baby is a boy?

8. Suppose that of all coins in the world one out of a million is two-headed and all the rest are fair coins. If a coin falls heads each time in 30 tosses, what is the probability that it is a two-headed coin?

9. In deriving the partition formula

$$P(B) = \sum_i P(A_i)P(B|A_i)$$

all we needed was $B = \bigcup_i BA_i$, that is, A_1, A_2, ... partition B, which may happen without $A_1, A_2, \ldots$ actually partitioning the whole sample set S (draw a Venn diagram or see Problems 1.2.16 and 1.2.17.) Also, the sequence A_1, A_2, ... need not be finite since P is completely additive (cf. Definition 1.5.1). In view of all this state a generalization of Proposition 2.2.2 and prove it.

10a. Given a probability space $(S, \mathcal{S}, P)$ prove the following "conditional product formula"

$$P(AB|D) = P(A|D)P(B|AD)$$

assuming that all conditional probabilities involved exist. Note that if we set $D = S$ we have the ordinary product formula. Hint: Use Lemma 2.2.1 and the ordinary product formula.

b. Derive the conditional partition formula (Proposition 2.2.4) from the conditional product formula above in much the same way as the partition formula (Proposition 2.2.2) was derived from the product formula.

2.3 STOCHASTIC INDEPENDENCE OF EVENTS

Given a pair of events A and B the conditional probability $P(B|A)$ may be larger or smaller than the (nonconditional) probability $P(B)$; it is therefore conceivable that for some special pair of events $P(B|A)$ may just equal $P(B)$. If $P(B|A) = P(B)$, this may be interpreted to mean that the occurrence of A in no way implies or precludes the occurrence of B; the event B is thus probabilistically independent from A. Accordingly we attempt the following definition.

2.3.1 DEFINITION

Given a probability space $(S, \mathscr{S}, P)$ and the two events A and B with $P(A) > 0$, we say B is *stochastically* (or probabilistically) *independent* from A if

$$P(B|A) = P(B)$$

If B is stochastically independent from A, we can easily show that A also is stochastically independent from B. Indeed, if B is independent from A, we have $P(B|A) = P(B)$ or

$$\frac{P(BA)}{P(A)} = P(B)$$

This is equivalent to

$$\frac{P(BA)}{P(B)} = P(A)$$

provided that $P(B) > 0$. In other words,

$$P(A|B) = P(A)$$

and A is stochastically independent from B; we have thus established the following proposition.

2.3.1 PROPOSITION

Given two events A and B with $P(A) \cdot P(B) > 0$, A is stochastically independent from B if and only if B is stochastically independent from A. In either case we have $P(AB) = P(A) \cdot P(B)$.

Since stochastic independence is a symmetric relation, we prefer the following definition to Definition 2.3.1.

2.3.2 DEFINITION

Given a probability space $(S, \mathscr{S}, P)$, events A and B are said to be stochastically independent if

$$P(AB) = P(A) \cdot P(B)$$

Note that this definition needs no assumption about $P(A)$ and $P(B)$; also note that a definition like this makes it necessary for $\mathscr{S}$ to be a Boolean algebra so that $P(AB)$ may be considered.

Although the concept of stochastic independence of events is defined within the framework of a probability space, in reality we often recognize stochastic independence of events associated with a certain experiment even before any probability space is conceived in detail. If two coins are tossed, we hardly expect an event pertaining to the first coin to imply or preclude in any way an event pertaining to the second coin. If the first coin falls heads, this in no way influences the probability of the second coin falling heads, or tails for that matter. This is so because the two coins do not communicate with each other. Likewise if a coin is tossed repeatedly, we hardly expect an event pertaining to the ith toss to imply or preclude in any way an event pertaining to the jth toss for $j \neq i$. If a brand-new coin falls heads the first 20 times, the probability that it falls heads on the 21st toss is still $\frac{1}{2}$. This is so because the coin, unlike us, has no memories. Incidentally, it is perhaps wiser to bet on heads on the 21st toss after 20 consecutive heads since we may suspect that the coin is biased. At any rate, there is no logical justification in betting on tails though, in practice, one may be tempted to bet on tails because of a false conclusion from the fact that the a priori (or nonconditional) probability of 21 consecutive heads for a brand-new coin is extremely small. In constructing a probability space $(S, \mathscr{S}, P)$ for an experiment, we must define P in such a way that for two events A and B, which we expect beforehand to be stochastically independent, we have $P(AB) = P(A) \cdot P(B)$. In practice, such an equation helps in completing the definition of P since after $P(A)$ and $P(B)$ are specified we need only recognize that A and B are independent to determine $P(AB)$.

Example 1 Two dice are separately known (or assumed) to have probabilities p' and p'' of giving a 6. If these two dice are thrown together, what is the probability of their giving a double-6.

Solution In the experiment of throwing these two dice let A be the event that the first die gives a 6 and B the event that the second die gives a 6, with $P(A) = p'$ and $P(B) = p''$. This is the only sensible thing to do. Now the event of a double-6 is represented by AB and since A and B are plainly stochastically independent, we should have

$$P(AB) = P(A) \cdot P(B) = p'p''$$

Example 2 The probability space that we usually associate with the experiment of tossing a pair of ordinary dice consists of 36 equally likely outcomes. Examine whether this model bears out the stochastic independence of, say, A, the event that the first die gives a 6 and B, the event that the second die gives a 6.

Solution In checking $P(AB) = P(A) \cdot P(B)$, we need only observe that $A = \{(6, 1), (6, 2), \ldots, (6, 6)\}$ so that $P(A) = \frac{6}{36} = \frac{1}{6}$; $B = \{(1, 6), (2, 6), \ldots, (6, 6)\}$ so that $P(B) = \frac{6}{36} = \frac{1}{6}$, and $AB = \{(6, 6)\}$ so that $P(AB) = \frac{1}{36}$.

If two events are not obviously stochastically independent, then an assumption about the probability space may settle the issue.

Example 3 Let A be the event that in a family of three children at most one is a girl, and B the event that the three children consist of both boys and girls. Determine within the framework of a probability space whether A and B are independent.

Solution Assuming the classical probability space in which the sample space consists of eight equally likely outcomes:

$$bbb, bbg, bgb, gbb, bgg, gbg, ggb, ggg$$

we see that

$$P(A) = \tfrac{4}{8} = \tfrac{1}{2}$$

and that

$$P(A \mid B) = \tfrac{3}{6} = \tfrac{1}{2}$$

Hence A is stochastically independent from B, and we see that A and B are independent events. It should be noted that A and B may not be independent if families of larger sizes are considered.

Example 4 If A and B are independent, does it follow that A and B' (B complement) are also independent?

Solution The occurrence of B is equivalent to nonoccurrence of B' and vice versa. If the occurrence of A made the occurrence of B' more (or less) likely, the occurrence of A would make the occurrence of B less (or more) likely, but this cannot be since A and B are independent. We can show this formally as follows (draw a Venn diagram):

$$P(AB') = P(A - AB) = P(A) - P(AB)$$
$$= P(A) - P(A) \cdot P(B)$$
$$= P(A)[1 - P(B)] = P(A) \cdot P(B')$$

If the independence of A and B implies the independence of A and B', likewise the independence of A and B' will imply the independence of A' and B'. Consequently the independence of A and B implies the independence of A' and B'.

Example 5 If A and B are independent, B and C are independent, and C and A are independent, in short A, B, C are pairwise independent, does it follow that AB and C are independent?

Solution The answer is no, as shown by the following example: In the experiment of tossing two ordinary dice, let A be the event of "odd toss" of the first die, B the event of "odd toss" of the second die, and C the event of "odd sum" of the two dice. First of all, clearly A and B are independent. Also it is not too difficult to see that B and C are independent or it can be checked formally by using the classical probability space of 36 equally likely outcomes. Likewise we see that C and A are independent. But AB and C cannot possibly be independent since the occurrence of AB means that the sum of two faces is even and this precludes the event C; as a matter of fact we have $P(C) = \frac{1}{2}$ while $P(C|AB) = 0$.

The preceding example shows that when A, B, C are pairwise independent, they may not be quite "thoroughly" independent since AB may not be independent from C, BC may not be independent from A, and so on. Yet, examples abound in which events involved are so thoroughly independent that any intersection event formed by some of them is invariably independent from any other event that did not take part in the formation of the intersection. For example if a coin is tossed 3 times, and A, B, C denote the event that the first toss, second toss, third toss is a head, then AB and C obviously are independent since the fact that the coin falls heads the first two times alters in no way the probability that the coin will fall heads the third time. If a coin is tossed 4 times and we failed to observe the first toss, then any information regarding the second, third, or fourth toss will not help in guessing the result of the first toss. This line of reasoning motivates the following definition.

2.3.3 DEFINITION

Events $A_1, A_2, A_3, \ldots$ are said to be *mutually independent* or to *form an independent class* if each of these events is independent from any intersection event formed by any number of other events. For example:

$$P(A_1) = P(A_1 | A_2 A_3 \cdots A_n)$$

or

$$P(A_1 A_2 \cdots A_n) = P(A_1) P(A_2 A_3 \cdots A_n)$$

Now from the last equation it is a matter of simple induction to show that

$$P(A_1 A_2 \cdots A_n) = P(A_1) P(A_2) \cdots P(A_n)$$

Accordingly we give the following more commonly accepted definition.

2.3.4 DEFINITION

A family of events are said to be *mutually independent* or to *form an independent class* if the probability of any intersection event is equal to the product of probabilities of events forming this intersection.

It can be easily shown that two intersection events formed by distinct events from an independent class are independent. It is also reassuring to see the following example.

Example 6 If A, B, C are from an independent class, show that $A \cup B$ and C are independent.

Solution With the aid of a Venn diagram we see that

$$P[(A \cup B)C] = P(AC \cup BC) = P(AC) + P(BC) - P(ABC)$$
$$= P(A)P(C) + P(B)P(C) - P(AB)P(C)$$
$$= [P(A) + P(B) - P(AB)]P(C)$$
$$= P(A \cup B)P(C)$$

2.3 PROBLEMS

1. Two ordinary dice are thrown. If A is the event that the total is not more than 10, and B is the event that the total is odd, are A and B independent?

2. Two coins with probabilities p_1 and p_2 of falling heads are tossed. Let A be the event that both coins fall heads, B the event that both coins fall tails, and C the event that one falls heads and the other tails. Can the values of p_1 and p_2 be so chosen that A, B, C become equally likely events?

3. Assuming that all permutations of three letters a, b, c are equally likely, determine whether the event that a precedes b and the event that b precedes c are independent.

4. A stick is broken into three pieces at random. If A is the event that the middle piece is more than half length, and B is the event that the left piece is longer than the right piece, are A and B independent? If C is the event that the left piece is more than a quarter length, are A and C independent?

5. Finite Bernoulli probability spaces. For the experiment of tossing n times a coin with probability p of falling heads, let the sample set S consist of all sequences of length n in two letters, h and t.

   a. If E_i is the subset of S consisting of all sequences whose ith letter is h, what event is represented by E_i?

   b. Let $\mathscr{S}$ be the family of all subsets of S. Show that each subset of S can be represented in terms of the basic events E_i. Hint: First consider the singleton subsets of S.

   c. If we set $P(E_i) = p$ for each $i = 1, 2, \ldots, n$, and assume that the events E_i are mutually independent, does this determine the probability of every elementary event? We might call the resulting $(S, \mathscr{S}, P)$ the (n, p) *Bernoulli probability space*.

   d. Use the (n, p) Bernoulli probability space as the model to calculate the probability of the event A_r that a p-coin falls heads exactly r times when tossed n times.

   e. A coin with probability $\frac{1}{3}$ of falling heads is tossed 5 times. What is the probability that it falls heads twice in succession before it has fallen tails twice in succession? Caution: Do not use the classical probability formula, since outcomes are not equally likely.

6. Five random digits are drawn. Use the Bernoulli model to determine the probability of exactly two of them being 0; of two or more of them being 9. For numerical values of $C_r^n p^r (1 - p)^{n-r}$ a binomial probability table is needed.

7. A machine consisting of n independent parts fails if any one of these n parts fails. If each part has the probability q of failing on any day, what is the probability of the machine failing on any day? Do the same problem assuming that the machine fails only if all n parts fail.

8. If the probability of each rifleman scoring a hit is $p = 0.03$, how many riflemen are needed to ensure that the probability of hitting the target is at least $P = 0.99$?

9. The normal rate of infection of a certain disease is known to be about 1 out of 4 exposed. To test the effect of a newly discovered vaccine 10 animals were exposed after the vaccination and only 2 were subsequently found to be infected. If the vaccine is totally ineffective, what is the probability of 2 infections out of 10 exposed? Is it advisable to recommend the vaccine as effective?

10. In a coin-tossing experiment X bets on one head, but Y believing that the coin is strongly biased toward tails offers to bet on two consecutive tails. How biased must the coin be in order for the two bettors to have equal chances of winning?

11. A thick coin falls heads, tails, or "sides" with probabilities p, q, and g where $p + q + g = 1$. Use an appropriate probability space to calculate the probability for the event $A_{r,s}$ of exactly r heads and s tails in n tosses of the coin. Can you construct a probability space for the experiment of tossing a loaded die n times? Hint: See Problem 1.4.3.

12. Show that two mutually exclusive events can never be independent unless one of them is an almost impossible event (having probability 0).

13. If A and C are independent and B and C are independent, show that a sufficient condition for $A \cup B$ and C to be independent is $AB = \phi$. Why is this condition not necessary? (Cf. Section 2.3, Example 6.)

14. Show that the sure event and the impossible event each taken with any other event form an independent pair of events. Show that an independent class of events remains independent with an addition or deletion of the sure event or the impossible event and in spite of a replacement of any event by its complementary event.

15. Show that events E_1, E_2, E_3, ... form an independent class if each event E_i is independent from any intersection event formed by any number of events preceding E_i.

2.4 BERNOULLI PROBABILITY SPACES

A mathematical model may be constructed to deal with problems at hand, but it may turn out useful in dealing with other problems also. Of course the initial model may have to be modified whenever necessary. In this section we shall construct Bernoulli probability spaces, which will be modified later to create Markov probability spaces. Let us begin by considering the following betting problem.

Example 1 In tossing a coin repeatedly, X bets on heads while B, believing that the coin is biased toward tails, offers to bet on two consecutive tails. If the coin turns out to be unbiased after all, what is the probability of X winning the bet? Of Y winning the bet?

Since clearly the contest will be over in two tosses, there is no possibility of a tie, and the probabilities of the two events can be easily determined. However, if X bets on two consecutive heads and Y bets on three consecutive tails, there is a definite possibility of a tie, and the probabilities of X winning or Y winning may not be so easy to determine. It is for problems such as these that we shall construct a Bernoulli probability space as our model. We proceed as follows:

First of all, since the tossing may have to go on beyond any preconceived length, we have no choice but to let our sample set consist of all infinite sequences in two letters, say, h (for heads) and t (for tails). Within this sample set S an elementary event would be a subset of S consisting of just one such infinite sequence. We note however that such a mathematically conceived elementary event has no counterpart in the real world since no one can actually finish tossing a coin infinitely many times. However, this need not disturb us as long as S contains enough subsets to represent real events that are of interest to us. For example, an event such as "heads appearing for the first time after seven consecutive tails" may be represented by that subset of S consisting of all infinite sequences which begin with seven consecutive t's followed immediately by an h.

Having agreed on the sample set S for the Bernoulli probability space $(S, \mathscr{S}, P)$, we must now decide what events to consider, that is, which subsets of S to include in $\mathscr{S}$. Although one may wish to declare simply that $\mathscr{S}$ includes all subsets of S, there is a serious mathematical difficulty involved in doing this. The fact is, if we include too many events in $\mathscr{S}$, we may not be able to define the probability of each and every event in $\mathscr{S}$ in a meaningful way: for instance, if we should assign probability to each event in $\mathscr{S}$ in such a way that for some sequence of pairwise disjoint events A_i the complete additivity of P fails, that is, $P(\bigcup_i A_i) \neq \sum_i P(A_i)$, then such a probability model would be hard to accept since we expect probabilities of disjoint events to add up just as we expect areas of disjoint geometric figures to add up whenever they are pieced together. It has actually been shown in measure theory that for too inclusive an $\mathscr{S}$ there may be no completely additive P. Therefore, the only sensible thing to do is not to take too large an $\mathscr{S}$. Evidently there are what we might call pathological events represented by some unusual subsets of S for which we simply cannot consider probability values consistent with those we assign readily to basic events. In measure theory such subsets of S are called *nonmeasurable*.

We shall let $\mathscr{S}$ include (1) those subsets of S which represent events of primary interest to us and in addition (2) those subsets which augment those in (1) to barely make $\mathscr{S}$ a sigma algebra. Now the events of primary interest to us are E_i, the event that the coin falls heads on the ith toss for $i = 1, 2, 3, \ldots$ Each E_i is represented by (or identified with) the subset of S consisting of all infinite sequences in which the ith letter is h, that is

$$E_i = \{\omega \in S : \omega_i = h\}$$

where ω_i is the ith letter of the infinite sequence ω. Next, we augment the collection of primary subsets E_i by taking the smallest sigma algebra of subsets of S that includes all the E_i. Such a sigma algebra, which can be shown to exist (see Problem 2.4.1), is then our choice of $\mathscr{S}$ for the Bernoulli probability space $(S, \mathscr{S}, P)$.

It remains to define P, a set function satisfying the axioms of probability measure [see (i) and (ii) of Definition 1.5.1]. The definition of P is based on the following compelling assumptions:

(1) Since $E_i \cup E_i^c = S$ for each $i = 1, 2, 3, \ldots$, that is, the coin either falls heads or tails at the ith toss, we set for each i

$$P(E_i) = p \quad \text{and} \quad P(E_i^c) = 1 - p$$

where p is the presumed probability of the coin falling heads in a single toss.

(2) Since the outcome of each toss in no way influences the outcome of any other toss, that is, the events E_i form an independent class, we set for each $n = 1$, 2, 3, $\ldots$ (see Definition 2.3.4)

$$P(E_{i_1} E_{i_2} \cdots E_{i_n}) = p^n$$

where $i_1, i_2, \ldots, i_n$ are n distinct values of $i = 1, 2, 3, \ldots$. Also,

$$P(E_{i_1}^c E_{i_2}^c \cdots E_{i_n}^c) = (1 - p)^n$$

and in general

$$P(E_{i_1}^\delta E_{i_2}^\delta \cdots E_{i_n}^\delta) = p^r (1 - p)^{n-r}$$

where r is the number of times in which δ, representing either void or c, is void.

We have defined P on the subclass $\mathscr{S}_0$ of $\mathscr{S}$ consisting of all events that are intersection events of the primary events E_i and their complement events E_i^c. Actually the subclass $\mathscr{S}_0$ has the structure of a semialgebra. In general, a class of sets is said to have the structure of a *semialgebra* if the intersection of any pair of sets in the class is again in the class and the complement of any set in the class can be expressed as a disjoint union of finitely many sets in the class. The reason why we point out that $\mathscr{S}_0$ is a semialgebra is to make use of the following result from measure theory (see Minibibliography: Royden).

Extension Principle of Measure

If P is a set function defined on a semialgebra $\mathscr{S}_0$ of subsets of S not in violation of axioms of (probability) measure, then P can be extended in a unique way to cover the smallest sigma algebra $\mathscr{S}$ containing $\mathscr{S}_0$ so that P on $\mathscr{S}$ satisfies all the axioms of (probability) measure.

For us this means that we are guaranteed of a unique definition of P over the entire $\mathscr{S}$ such that the effect of P on $\mathscr{S}_0$ is exactly as stipulated in (1) and (2) above, and that P has all the required properties of probability measure, including the complete additivity property. The construction of Bernoulli probability space just completed will substantiate the following formal definition. We point out incidentally that without substantiation a formal definition in mathematics runs the risk of describing something mathematically nonexistent.

2.4.1 DEFINITION

By a Bernoulli probability space we mean a triple $(S, \mathscr{S}, P)$ in which S is the set of all infinite sequences in 0 and 1, $\mathscr{S}$ is the smallest sigma algebra of subsets of S containing the E_i for $i = 1, 2, 3, \ldots$, where E_i is the subset of S consisting of all sequences in which the ith term is 1; and P is a probability measure on $\mathscr{S}$ satisfying:

(1) $P(E_i) = p$ for $i = 1, 2, 3, \ldots$ where $0 < p < 1$ is a fixed constant.
(2) $P(E_{i_1} E_{i_2} \cdots E_{i_n}) = p^n$ for any finite set of integers $0 < i_1 < i_2 < \cdots < i_n$.

As an immediate consequence of Definition 2.4.1 we have the following proposition.

2.4.1 PROPOSITION

In the Bernoulli probability space $(S, \mathscr{S}, P)$ we have for any intersection event of finitely many events E_i or E_i^c the following simple probability formula

$$P(E_{i_1}^\delta E_{i_2}^\delta \cdots E_{i_n}^\delta) = p^r(1 - p)^{n-r}$$

where δ represents either c or void, and r is the number of times in which δ represents void.

PROOF

From (1) and (2) of Definition 2.4.1 we have

$$P(E_{i_1} E_{i_2} \cdots E_{i_n}) = P(E_{i_1}) P(E_{i_2}) \cdots P(E_{i_n})$$

Thus, the family of events E_i forms an independent class, but then so does the family of these events with any number of them replaced by their complement events (see Problem 2.3.14), hence

$$P(E_{i_1}^\delta E_{i_2}^\delta \cdots E_{i_n}^\delta) = P(E_{i_1}^\delta) P(E_{i_2}^\delta) \cdots P(E_{i_n}^\delta)$$

It remains to show that $P(E_i^c) = 1 - p$, but this is trivial since $E_i \cup E_i^c = S$ and $P(E_i^c) = P(S) - P(E_i) = 1 - p$. This completes the proof.

In our definition of Bernoulli probability space only the value of p is arbitrary. Therefore, a unique Bernoulli probability space is specified as soon as a specific value of p is chosen. This choice is the only subjective element in the model building of Bernoulli probability space.

Example 2 A coin falls heads about once out of every three tosses. If this coin is tossed repeatedly, what should be the probability of its falling heads exactly twice during the first six tosses?

Solution We use the Bernoulli probability space with $p = \frac{1}{3}$ as our model to determine the probability of the event A in question. A can be split into $C_2^6 = 15$ subsets, for example $E_1 E_2 E_3^c E_4^c E_5^c$, each of which has probability $(\frac{1}{3})^2 (\frac{2}{3})^4$ by Proposition 1. Consequently,

$$P(A) = C_2^6 (\tfrac{1}{3})^2 (\tfrac{2}{3})^4 = \tfrac{80}{243}$$

Example 3 A business executive has a large pool of advisers each of whom offers correct advice about 60% of the time. In making an important decision the executive usually consults an odd number of advisers and follows the majority opinion. Calculate the probability p_k that the executive makes a correct decision if he consults k advisers for $k = 5, 7$, and 19. Is the executive wasting his time by increasing k?

Solution If the executive consults 5 advisers, he will arrive at a correct decision only if three or more advisers offer correct advices, and so on. Simulating each advisor by a coin with probability $p = 0.6$ of falling heads, and using Bernoulli probability space with $p = 0.6$ as model, we calculate p_k as follows:

$$p_5 = \sum_{r=3}^{5} P(A_{5,r}) = \sum_{r=3}^{5} C_r^5 (0.6)^r (0.4)^{5-r} = 0.6826$$

$$p_7 = \sum_{r=4}^{7} P(A_{7,r}) = \sum_{r=4}^{7} C_r^7 (0.6)^r (0.4)^{7-r} = 0.7102$$

$$p_{19} = \sum_{r=10}^{19} P(A_{19,r}) = \sum_{r=10}^{19} C_r^{19} (0.6)^r (0.4)^{19-r} = 0.8139$$

where $A_{n,r}$ denotes the event of r heads in n tosses so that

$$P(A_{n,r}) = C_r^n p^r (1-p)^{n-r}$$

We see that p_k seems to increase as k increases. In fact it can be shown (cf. the weak law of large numbers, Theorem 6.1.1) that p_k approaches 1 as k approaches ∞ as long as $p > 0.5$. If $p = 0.5$, then $p_k = 0.5$ for all k and nothing is gained by increasing k.

Example 4 A coin with probability p of falling heads is tossed repeatedly. By-standers X and Y respectively bet on α consecutive heads and β consecutive tails; either event occurring first determines the winner. Calculate the probabilities of (1) X winning the bet, (2) Y winning the bet, and (3) a tie.

Solution Let A be the event that α consecutive heads appear before β consecutive tails have a chance to, B the event that β consecutive tails appear before α consecutive heads have a chance to, and C the complement event of A and B. We consider A, B, C as represented in the Bernoulli probability space $(S, \mathscr{S}, P)$ and determine $P(A), P(B), P(C)$.

To determine $P(A)$ consider first the following two "key" conditional probabilities:

$$u = P(A|E_1)$$
$$v = P(A|E_1^c)$$

Many other conditional probabilities are ultimately reduced to u and v. Note for instance that

$$P(A|E_1 E_2^c) = P(A|E_1^c) = v$$
$$P(A|E_1^c E_2) = P(A|E_1) = u$$

since whenever the coin falls tails X has to start counting the heads all over again, and so on. Note also that

$$P(A|E_1) \leqq P(A|E_1 E_2) \leqq \cdots \leqq P(A|E_1 E_2 \cdots E_\alpha) = 1$$
$$P(A|E_1^c) \geqq P(A|E_1^c E_2^c) \geqq \cdots \geqq P(A|E_1^c E_2^c \cdots E_\beta^c) = 0$$

since X comes closer to winning as more heads appear consecutively, and so on.

According to the partition formula (Proposition 2.2.2) we have

$$P(A) = P(E_1)P(A|E_1) + P(E_1^c)P(A|E_1^c)$$
$$= pu + qv$$

where $q = 1 - p$. Hence $P(A)$ is determined as soon as u and v are. To determine u and v we shall find two equations satisfied by them, and then solve these equations.

Now, referring to the first of the two chain inequalities above, we shall relate each term to the next, thus ultimately relating the first, $u = P(A|E_1)$, to the last, $P(A|E_1 E_2 \cdots E_\alpha) = 1$. To do this we use the conditional partition formula (Proposition 2.2.4) to obtain

$$P(A|E_1 E_2 \cdots E_r) = P(E_{r+1}|E_1 E_2 \cdots E_r)P(A|E_1 E_2 \cdots E_r E_{r+1})$$
$$+ P(E_{r+1}^c|E_1 E_2 \cdots E_r)P(A|E_1 E_2 \cdots E_r E_{r+1}^c)$$
$$= P(E_{r+1})P(A|E_1 E_2 \cdots E_{r+1}) + P(E_{r+1}^c)P(A|E_1^c)$$
$$= pP(A|E_1 E_2 \cdots E_{r+1}) + qv$$

Thus,

$$u = P(A \mid E_1) = pP(A \mid E_1E_2) + qv$$
$$= p[pP(A \mid E_1E_2 E_3) + qv] + qv$$
$$= \cdots$$

and routine calculations reveal that

$$u = p^{\alpha-1} + (1 - p^{\alpha-1})v$$

Likewise, working on the second chain of inequalities, we see

$$v = (1 - q^{\beta-1})u$$

We solve the above two equations for u and v to obtain

$$u = \frac{p^{\alpha-1}}{1 - (1 - p^{\alpha-1})(1 - q^{\beta-1})}$$

$$v = \frac{p^{\alpha-1}(1 - q^{\beta-1})}{1 - (1 - p^{\alpha-1})(1 - q^{\beta-1})}$$

Consequently,

$$P(A) = \frac{p^{\alpha-1}(1 - q^{\beta})}{1 - (1 - p^{\alpha-1})(1 - q^{\beta-1})}$$

For example, if $p = \frac{1}{2}$, $\alpha = 2$, and $\beta = 3$, we have $P(A) = 0.7$.

To find $P(B)$ we merely interchange p and q, and α and β, in the above expression for $P(A)$. Thus,

$$P(B) = \frac{q^{\beta-1}(1 - p^{\alpha})}{1 - (1 - p^{\alpha-1})(1 - q^{\beta-1})}$$

Again, if $p = \frac{1}{2}$, $\alpha = 2$, and $\beta = 3$, we have $P(B) = 0.3$.

As for $P(C)$, from $P(C) = 1 - P(A \cup B)$ and $P(A) + P(B) = 1$, we obtain

$$P(C) = 0$$

In other words, the probability of a tie is 0 according to our model.

2.4 PROBLEMS

1 a. If $\mathcal{B}_1$ and $\mathcal{B}_2$ are two sigma algebras of subsets of S, show that their intersection $\mathcal{B}_1 \cap \mathcal{B}_2$ is also a sigma algebra of subsets of S. More generally show that if $\{\mathcal{B}_i\}_i$ is a collection of sigma algebras of subsets of S their intersection

$$\bigcap_i \mathcal{B}_i = \{B \subset S \colon \mathcal{B} \in \mathcal{B}_i \quad \text{for all } i\}$$

is still a sigma algebra of subsets of S.

b. Show that given an arbitrary family $\mathscr{E}$ of subsets of S there exists a smallest (in the sense of being contained in every other such) sigma algebra $\mathscr{S}$ of subsets of S which includes the family $\mathscr{E}$. $\mathscr{S}$ is called the sigma algebra *generated* by $\mathscr{E}$, and members of $\mathscr{E}$ may be called the *primary* members of $\mathscr{S}$.

2. Let $\mathscr{E}$ be a family of subsets of S, and let $\mathscr{S}_0$ consist of all finite intersections of subsets belonging to $\mathscr{E}$, so that clearly $\mathscr{E} \subset \mathscr{S}_0$. Show however that the sigma algebra generated by $\mathscr{S}_0$ is no different from the sigma algebra generated by $\mathscr{E}$.

3 a. Show that the family $\mathscr{S}_0$ of all finite intersections $E_{i_1} E_{i_2} \cdots E_{i_n}$ of primary events E_i in a Bernoulli probability space is a semialgebra (cf. page 46).
   b. Show that the family $\mathscr{S}_1$ of all finite unions of events belonging to $\mathscr{S}_0$ is a Boolean algebra (cf. Definition 1.3.2).
   c. Show that every sigma algebra is automatically a Boolean algebra, and every Boolean algebra is automatically a semialgebra, but that the converses are not true.

4. A probability space in which all elementary events have 0 probability is called a *continuous probability space*. A probability space in which the probabilities assigned to elementary events add up to 1 is called a *discrete probability space*.
   a. Show that every finite probability space is a discrete probability space, but not vice versa.
   b. Show by examples that a probability space may be neither discrete nor continuous. Such a probability space may be called a *mixed* probability space.

5. In a Bernoulli probability space $(S, \mathscr{S}, P)$ show (a) that every elementary event is included in $\mathscr{S}$ and (b) that if E is an elementary event, $P(E) = 0$ and that therefore a Bernoulli probability space is a continuous probability space.

2.5 MARKOV PROBABILITY SPACES

We shall now construct Markov probability spaces, of which the Bernoulli probability spaces may be regarded as a very special case. Following the pattern set by the preceding section, we shall give the axiomatic definition of Markov probability spaces (Definition 2.5.1) only after the actual constructions.

For a Markov probability space $(S, \mathscr{S}, P)$, the sample set S consists of infinite sequences in K integers, 1, 2, ..., K. $\mathscr{S}$ is the sigma algebra of subsets of S generated by certain "primary" events. These are the events E_i^j for $j = 1, 2, 3, \ldots,$ K, and $i = 0, 1, 2, \ldots,$ where E_i^j is the set of all infinite sequences whose ith term is the integer j. As for the construction of the probability measure P, again in view of the extension principle of measure we shall only specify its values on a certain semialgebra $\mathscr{S}_0$, which generates precisely the sigma algebra $\mathscr{S}$. However, lest our definition of P on $\mathscr{S}_0$ appear too arbitrary, we shall first consider a typical problem which a Markov probability space is designed to handle.

Example 1 X has six sons whom we shall label by $j = 1, 2, 3, \ldots, 6$. At the end of the day X casts a die to decide which of the six sons he will visit the next day. The six dice placed at the six sons' houses are all loaded, presumably differently. Let us suppose that the jth die is loaded according to the numbers p_{jk} with

$$\sum_{k=1}^{6} p_{jk} = 1$$

so that altogether we have a 6×6 array (hence a matrix) of positive real numbers

$$\begin{bmatrix} p_{11} & p_{12} & \cdots & \cdots & \cdots & p_{16} \\ p_{21} & p_{22} & \cdots & \cdots & \cdots & p_{26} \\ \cdots & \cdots & \cdots & \cdots & \cdots & \cdots \\ \cdots & \cdots & \cdots & \cdots & \cdots & \cdots \\ \cdots & \cdots & \cdots & \cdots & \cdots & \cdots \\ p_{61} & \cdots & \cdots & \cdots & \cdots & p_{66} \end{bmatrix}$$

whose six rows all have row sums equal to 1, thus describing the stochastic natures of the six dice. If on some particular day ($i = 0$) when we start observing X, he is found at $j = 6$, what is the probability that X will be at, say, $j = 1$ two days later ($i = 2$)? or what is the probability that X will be at $j = 6, 5, 4, 3$ in that order during the first four days?

Solution It is immediately clear that the sequence of visits of X is best described by infinite sequences in six integers, that is, the sample set S with $K = 6$ in our proposed Markov probability space $(S, \mathscr{S}, P)$. The event that X is at 6 when $i = 0$ is then represented by E_0^6; likewise the event that X is at 1 when $i = 2$ is represented by E_2^1; and the event that X is at 6, 5, 4, 3, when $i = 0, 1, 2, 3$ is represented by the intersection $E_0^6 E_1^5 E_2^4 E_3^3$ of four primary events. Now quite obviously the probabilities of these events are dictated by p_{jk}. First of all, trivially

$$P(E_0^6) = 1$$

since X is found at 6 when $i = 0$. Next, following the partition formula (Proposition 2.2.2), since

$$E_2^1 = (E_1^1 \cup E_1^2 \cup \cdots \cup E_1^6)E_2^1$$

we obtain

$$P(E_2^1) = \sum_{j=1}^{6} P(E_1^j)P(E_2^1 | E_1^j)$$

$$= \sum_{j=1}^{6} p_{6j} p_{j1}$$

where $P(E_1^j) = p_{6j}$ since the sixth die is cast at $j = 6$ when $i = 0$ to determine the next position j for $i = 1$, and likewise $P(E_2^1 | E_1^j) = p_{j1}$ since the jth die is cast at the jth position, and so on.

As for the event $E_0^6 E_1^5 E_2^4 E_3^3$, the extended product formula (Proposition 2.2.1) would dictate

$$P(E_0^6 E_1^5 E_2^4 E_3^3) = P(E_0^6)P(E_1^5 | E_0^6)P(E_2^4 | E_0^6 E_1^5)P(E_3^3 | E_0^6 E_1^5 E_2^4)$$

but now the following important observation comes in to simplify the calculation

$$P(E_3^3 | E_0^6 E_1^5 E_2^4) = P(E_3^3 | E_2^4)$$

since the probability of X being at 3 when $i = 3$ depends only on the fourth die, that is only on X taking position 4 at $i = 2$, and not on where X had been prior to $i = 2$.

Similarly we have

$$P(E_2^4 | E_0^6 E_1^5) = P(E_2^4 | E_1^5)$$

and so on; consequently we have

$$P(E_0^6 E_1^5 E_2^4 E_3^3) = P(E_0^6)P(E_1^5 | E_0^6)P(E_2^4 | E_1^5)P(E_3^3 | E_2^4)$$

$$= p_{65} p_{54} p_{43}$$

and again the p_{jk} dictate the probability of the event.

It is now clear how we should define the probability measure P in general for a Markov probability space $(S, \mathscr{S}, P)$. Let $\mathscr{S}_0$ be the semialgebra of subsets of S consisting of intersections of the form

$$E_0^{j_0} E_1^{j_1} E_2^{j_2} \cdots E_n^{j_n}$$

where j_i are integers among $1, 2, 3, \ldots, K$. We shall not burden ourselves with the routine proof of the fact that $\mathscr{S}_0$ is indeed a semialgebra. For each such intersection event we now set

$$P(E_0^{j_0} E_1^{j_1} E_2^{j_2} \cdots E_n^{j_n}) = q_{j_0} p_{j_0 j_1} p_{j_1 j_2} \cdots p_{j_{n-1} j_n}$$

where the p_{jk} come from some $K \times K$ array (matrix) of nonnegative real numbers whose row sums are all equal to 1, and $q_{j_0} = P(E_0^{j_0})$, which may in general be less than 1, is one of the nonnegative numbers $q_1, q_2, \ldots, q_k$ totaling 1. Again we shall not burden ourselves with the proof that P so defined on $\mathscr{S}_0$ does not violate the axioms of probability measure. Thus, in view of the extension principle of measure our construction of P and hence of $(S, \mathscr{S}, P)$ is complete. We state the following formal definition.

2.5.1 DEFINITION

By a Markov probability space we mean a probability space $(S, \mathscr{S}, P)$ in which S consists of all infinite sequences of K integers $1, 2, \ldots, K$; $\mathscr{S}$ is the sigma algebra of subsets of S generated by the subsets E_i^j of S for $j = 1, 2, \ldots, K$ and $i = 0$,

1, 2, 3, . . . , where E_i^j is the set of all sequences whose ith term is j; and P satisfies the following conditions:

(1) For any $i = 0, 1, 2, 3, \ldots,$

$$P(E_{i+1}^k \mid DE_i^j) = p_{jk}$$

where D is any member of the Boolean algebra generated by the primary events E_h^j for $h < i$ and $j = 1, 2, 3, \ldots, K$, and the $p_{jk} \geq 0$ constitute a $K \times K$ non-negative matrix

$$(p_{jk}) = M$$

whose row sums are all equal to 1, that is, $\sum_{k=1}^{K} p_{jk} = 1$ for $j = 1, 2, \ldots, K$.

(2) For $j = 1, 2, \ldots, K$,

$$P(E_0^j) = q_j$$

where $q_j \geq 0$ constitute a nonnegative row vector

$$(q_1, q_2, \ldots, q_K) = \mathbf{q}$$

whose row sum is equal to 1, that is, $\sum_{k=1}^{K} q_j = 1$. In particular, one may have, say $q_1 = q_2 = \cdots = q_{K-1} = 0$ and $q_K = 1$.

From Definition 2.5.1 it is clear that a specific Markov probability space is chosen by specifying (1) the value of a positive integer K (henceforth referred to as the number of *states*), (2) a $K \times K$ nonnegative matrix M with row sums equal to 1 (henceforth referred to as the *matrix of transition probabilities* or simply *transition matrix*), and (3) a nonnegative K-dimensional row vector $\mathbf{q}$ with row sum equal to 1 (henceforth referred to as the *initial probability vector*). Since the number of states K is subsumed by the size of the transition matrix M, a specific Markov probability space is given by a joint indication $(M, \mathbf{q})$ of transition matrix M and the initial probability vector $\mathbf{q}$.

Condition (1) of Definition 2.5.1 has far-reaching implications. First of all, since $P(E_{i+1}^k \mid DE_i^j)$ remain unchanged for different choices of D, allowing in particular $D = S$ [cf. Problem 1.5.12(iii)] and since $SE_i^j = E_i^j$, we obtain

$$P(E_{i+1}^k \mid DE_i^j) = P(E_{i+1}^k \mid E_i^j) \tag{1a}$$

Thus the conditional probability of E_{i+1}^k given DE_i^j depends not so much on DE_i^j as on E_i^j. More generally, we can show that condition (1) implies that the conditional probability of E_{i+n}^k given DE_i^j depends not so much on DE_i^j as on E_i^j for all $n = 1, 2, 3, \ldots$.

2.5.1 PROPOSITION

Given a Markov probability space $(S, \mathscr{S}, P)$ we have

$$P(E_{i+n}^k \mid DE_i^j) = P(E_{i+n}^k \mid E_i^j) \qquad n = 1, 2, 3, \ldots$$

for any member D of the Boolean algebra generated by the primary events E_h^j for $h < i$ and $j = 1, 2, 3, \ldots, K$.

PROOF

We give an inductive proof. For $n = 1$, we have equation (1a). We make the inductive assumption

$$P(E_{i+n-1}^k | DE_i^j) = P(E_{i+n-1}^k | E_i^j)$$

Now,

$$P(E_{i+n}^k | DE_i^j) = \sum_{h=1}^{K} P(E_{i+n-1}^h | DE_i^j) P(E_{i+n}^k | DE_i^j E_{i+n-1}^h)$$

by the conditional partition formula (Proposition 2.2.4). In view of the inductive assumption and condition (1) this reduces to

$$\sum_{h=1}^{K} P(E_{i+n-1}^h | E_i^j) P(E_{i+n}^k | E_i^j E_{i+n-1}^h)$$

Using the conditional partition formula backward, we obtain

$$P(E_{i+n}^k | DE_i^j) = P(E_{i+n}^k | E_i^j)$$

completing the proof.

The transition matrix $M = (p_{jk})$ summarizes in a nutshell what might be called the one-step transition probabilities of the Markov probability space since

$$p_{jk} = P(E_{i+1}^k | E_i^j)$$

Note that these one-step transition probabilities going from i to $i + 1$ are independent of i. More generally we will now consider the n-step (or multiple-step) transition probabilities for $n = 1, 2, 3, \ldots$ and show that they too are independent of i (Proposition 2.5.2).

2.5.2 DEFINITION

We define the n-step transition probabilities by

$$p_{jk}^{(n)} = P(E_{i+n}^k | E_i^j)$$

and call the resulting matrix

$$M^{(n)} = (p_{jk}^{(n)})$$

the n-step transition matrix.

We hasten to add that strictly speaking the preceding definition should not be given until after we have shown that the $P(E_{i+n}^k | E_i)$ are independent of i. This we will do now.

2.5.2 PROPOSITION

For an arbitrarily fixed value of i, let

$$p_{jk}^{(n)} = P(E_{i+n}^k | E_i^j)$$

then the matrix $M^{(n)} = (p_{jk}^{(n)})$ is related to the one-step transition matrix M by

$$M^{(n)} = M^n$$

Thus, $M^{(n)}$ being the nth power of M, which is independent of i, must be independent of i also.

PROOF
The proof is by induction. For $n = 1$ the assertion is trivial. So we assume $M^{(n-1)} = M^{n-1}$ to prove $M^{(n)} = M^n$. Now,

$$p_{jk}^{(n)} = P(E_{i+n}^k | E_i^j)$$

$$= \sum_{h=1}^{K} P(E_{i+n-1}^h | E_i^j) P(E_{i+n}^k | E_i^j E_{i+n-1}^h)$$

$$= \sum_{h=1}^{K} p_{jh}^{(n-1)} p_{hk}$$

where we used the conditional partition formula (Proposition 2.2.4) for the second equality. We see that if we multiply the matrices M^{n-1} and M we obtain as the (j, k) entry exactly the last term of the above equation. Thus, the (j, k) entry of M^n is equal to the (j, k) entry of $M^{(n)}$, and hence $M^{(n)} = M^n$.

The formal definition of a Markov probability space tells us very little about actual calculations of probabilities of events. The multiple-step transition matrices will now enable us to deal with this particular aspect of Markov probability spaces. The following proposition is our counterpart to Proposition 2.4.1.

2.5.3 PROPOSITION
Given a Markov probability space $(S, \mathcal{S}, P)$ with P specified by $(M, \mathbf{q})$,

(a) for each i the probabilities of primary events $E_i^1, E_i^2, \ldots, E_i^K$ can be found collectively from the matrix multiplication

$$\mathbf{q}_i = \mathbf{q} M^i$$

where $\mathbf{q}_i$ is the row vector $[P(E_i^1), P(E_i^2), \ldots, P(E_i^K)]$.

(b) for any m values of i, $0 \leq i_1 < i_2 < \cdots < i_m$, and m values of $j, j_1, j_2, \ldots, j_m$, we have

$$P(E_{i_1}^{j_1} E_{i_2}^{j_2} \cdots E_{i_m}^{j_m}) = q_{i_1}^{j_1} p_{j_1 j_2}^{(i_2 - i_1)} \cdots p_{j_{m-1} j_m}^{(i_m - i_{m-1})}$$

where $q_{i_1}^{j_1}$ is the j_1th term of the row vector $\mathbf{q}_{i_1}$ [see (a)] and $p_{j_1 j_2}^{(i_2 - i_1)}$ is the (j_1, j_2) entry of the $(i_2 - i_1)$-step transition matrix $M^{i_2 - i_1}$, and so on.

Note that M and $\mathbf{q}$ thus dictate the probabilities of all events considered.

PROOF
(a) The proof is by induction. First we must show for $i = 1$ that $\mathbf{q}_1 = \mathbf{q} M$ holds. Now

$$\mathbf{q}_1 = [P(E_1^1), P(E_1^2), \ldots, P(E_1^K)]$$

but in view of the partition formula (Proposition 2.2.2)

$$P(E_1^j) = \sum_{h=1}^{K} P(E_0^h)P(E_1^j \mid E_0^h) = \sum_{h=1}^{K} q_h p_{hj}$$

In other words, the jth term of $\mathbf{q}_1$ is exactly the jth term of the matrix product $\mathbf{q}M$, hence $\mathbf{q}_1 = \mathbf{q}M$.

Next, we make the inductive assumption $\mathbf{q}_{i-1} = \mathbf{q}M^{i-1}$ to show $\mathbf{q}_i = \mathbf{q}M^i$. Now,

$$\mathbf{q}_i = [P(E_i^1), P(E_i^2), \ldots, P(E_i^K)]$$

but again in view of the partition formula we have

$$P(E_i^j) = \sum_{h=1}^{K} P(E_{i-1}^h)P(E_i^j \mid E_{i-1}^h) = \sum_{h=1}^{K} q_{i-1}^h p_{hj}$$

In other words, the jth term of $\mathbf{q}_i$ is exactly the jth term of the matrix product $\mathbf{q}_{i-1}M$, but by the inductive assumption $\mathbf{q}_{i-1}M = (\mathbf{q}M^{i-1})M = \mathbf{q}M^i$, hence $\mathbf{q}_i = \mathbf{q}M^i$.

(b) By the extended product formula (Proposition 2.2.1) we have

$$P(E_{i_1}^{j_1}E_{i_2}^{j_2} \cdots E_{i_m}^{j_m}) = P(E_{i_1}^{j_1})P(E_{i_2}^{j_1} \mid E_{i_1}^{j_1}) \cdots P(E_{i_m}^{j_m} \mid E_{i_1}^{j_1} \cdots E_{i_{m-1}}^{j_{m-1}})$$

but by Proposition 1 this reduces to

$$P(E_{i_1}^{j_1})P(E_{i_2}^{j_2} \mid E_{i_1}^{j_1}) \cdots P(E_{i_m}^{j_m} \mid E_{i_{m-1}}^{j_{m-1}})$$

which, in view of (a) and the multiple-step transition matrices, further reduces to the desired expression, completing the proof.

In the preceding proposition we considered the probabilities of the basic events E_i^1, E_i^2, $\ldots$, E_i^K collectively for each i. We now state formally the following definition.

2.5.3 DEFINITION
Given a Markov probability space $(S, \mathscr{S}, P)$ the row vector

$$\mathbf{q}_i = [P(E_i^1), P(E_i^2), \ldots, P(E_i^K)]$$

is called the *ith (stage) probability vector*.

We see that Proposition 2.5.3(a) simply says that the ith probability vector $\mathbf{q}_i$ can be derived from the initial probability vector $\mathbf{q}$ by a routine matrix multiplication involving the transition matrix M.

One of the important problems concerning Markov probability space is to investigate the nature of $\mathbf{q}_i$ for large values of i. The mathematical formulation of this is the investigation of existence as well as the determination of $\lim_{i \to \infty} \mathbf{q}_i$ for a given Markov probability space $(M, \mathbf{q})$. Before going into this problem let us see an example to acquire the feel for $\mathbf{q}_i$.

Example 2 In giving a long series of quizzes to a large class of students the instructor is interested in estimating the percentage of students to pass the fifth quiz. Assuming that a student passing a quiz will fail the next quiz with probability 0.2 and that a student failing a quiz will pass the next quiz with probability 0.1, and supposing that 60% of the class passed the preliminary (initial) quiz, calculate the probability that a student passes the fifth quiz.

Solution We use as our model the 2-state (pass or fail) Markov probability space $(M, \mathbf{q})$ where

$$M = \begin{bmatrix} p_{11} & p_{12} \\ p_{21} & p_{22} \end{bmatrix} = \begin{bmatrix} 0.8 & 0.2 \\ 0.1 & 0.9 \end{bmatrix}$$

and $\mathbf{q} = (0.6, 0.4)$. Our problem is to find the row vector $\mathbf{q}_5$ and read off the first component. Now since $\mathbf{q}_5 = \mathbf{q}M^5$ by Proposition 2.5.3(a), a routine matrix calculation gives us $\mathbf{q}_5 \doteq (0.38, 0.62)$. Hence a student will pass the fifth quiz with probability 0.38. In other words, about 38% of the class will pass the fifth quiz.

Example 3 Given the data of Example 2, does the probability of a student passing the ith quiz tends to a fixed constant as i gets larger?

Solution Continuing with the same model, we shall investigate $\lim_{i \to \infty} \mathbf{q}_i$. The routine matrix calculation may get us $\mathbf{q}_i$ for each individual i but not tell us anything about the tendency of $\mathbf{q}_i$. This is one of the reasons why we cannot always depend on routine calculations in mathematics. To proceed let us rewrite our initial probability vector as

$$\mathbf{q} = (p_0, q_0)$$

where $p_0 = 0.6$ and $q_0 = 0.4$, the ith probability vector as

$$\mathbf{q}_i = (p_i, q_i)$$

and the transition matrix as

$$M = \begin{bmatrix} 1 - \alpha & \alpha \\ \beta & 1 - \beta \end{bmatrix}$$

where $\alpha = 0.2$ and $\beta = 0.1$. Now let us see in what way each $\mathbf{q}_i$ depends on the preceding $\mathbf{q}_{i-1}$. Since $\mathbf{q}_i = \mathbf{q}M^i = (\mathbf{q}M^{i-1})M = \mathbf{q}_{i-1}M$,

$$(p_i, q_i) = (p_{i-1}, q_{i-1}) \begin{bmatrix} 1 - \alpha & \alpha \\ \beta & 1 - \beta \end{bmatrix}$$

Therefore,

$$p_i = (1 - \alpha)p_{i-1} + \beta q_{i-1}$$
$$= (1 - \alpha - \beta)p_{i-1} + \beta$$

since $q_{i-1} = 1 - p_{i-1}$. This recurrence formula gives us a way of ultimately relating p_i to p_0. In fact,

$$p_i = (1 - \alpha - \beta)^i p_0 + (1 - \alpha - \beta)^{i-1}\beta + \cdots + (1 - \alpha - \beta)\beta + \beta$$

$$= (1 - \alpha - \beta)^i p_0 + \frac{1 - (1 - \alpha - \beta)^i}{1 - (1 - \alpha - \beta)} \beta$$

so that for instance if $\alpha = 0.2$, $\beta = 0.1$, and $n = 5$, then $p_5 \doteq 0.38$. Now for i very large, $(1 - \alpha - \beta)^i$ is negligible, that is, we may write

$$p_i = \varepsilon p_0 + \frac{1 - \varepsilon}{\alpha + \beta} \beta$$

where ε approaches 0 as i approaches ∞. Consequently we see

$$\lim_{i \to \infty} p_i = \frac{\beta}{\alpha + \beta} = \frac{1}{3} \doteq 0.33$$

and

$$\lim_{i \to \infty} q_i = \frac{\alpha}{\alpha + \beta} = \frac{2}{3}$$

since $q_i = 1 - p_i$. Thus

$$\lim_{i \to \infty} q_i = \left[\frac{\beta}{\alpha + \beta}, \frac{\alpha}{\alpha + \beta} \right] = (\tfrac{1}{3}, \tfrac{2}{3})$$

We have just seen a Markov probability space $(M, \mathbf{q})$, in which the limiting probability vector $\lim_{i \to \infty} \mathbf{q}_i = \mathbf{q}_\infty$ exists. In fact $\mathbf{q}_\infty$ looked as if it had nothing to do with the initial probability vector $\mathbf{q}$. This leads us to the following questions. Given a Markov probability space $(M, \mathbf{q})$,

(a) When does $\mathbf{q}_\infty$ exist?
(b) When does $\mathbf{q}_\infty$ exist independently of $\mathbf{q}$?
(c) If $\mathbf{q}_\infty$ exists independently of $\mathbf{q}$, how does one find such $\mathbf{q}_\infty$ directly from M?

Before attempting to answer these questions we should see a simple example of $(M, \mathbf{q})$ in which $\mathbf{q}_\infty$ does not exist.

Example 4 Given $(M, \mathbf{q})$ with M the 2×2 matrix $\begin{bmatrix} 0 & 1 \\ 1 & 0 \end{bmatrix}$ and $\mathbf{q} = (1, 0)$, show that $\lim_{i \to \infty} \mathbf{q}_i$ does not exist.

Solution It is easy to see that $\mathbf{q}_i = (0, 1)$ or $(1, 0)$ according to whether i is odd or even. Since the $\mathbf{q}_i$ fluctuate perpetually between $\mathbf{v}_1 = (1, 0)$ and $\mathbf{v}_2 = (0, 1)$, $\mathbf{q}_i$ approaches no limit.

Next we see a trivial case in which $\mathbf{q}_\infty$ exists.

2.5.4 PROPOSITION

If a Markov probability space $(M, \mathbf{q})$ is such that $\mathbf{q}M = \mathbf{q}$, then $\lim_{i \to \infty} \mathbf{q}_i$ exists, in fact, $\mathbf{q}_\infty = \mathbf{q}$.

PROOF

Applying $\mathbf{q}M = \mathbf{q}$, i times, we see

$$\begin{aligned}
\mathbf{q}_i = \mathbf{q}M^i &= (\mathbf{q}M)M^{i-1} \\
&= \mathbf{q}M^{i-1} \\
&= \mathbf{q}M^{i-2} \\
&\;\;\vdots \\
&= \mathbf{q}
\end{aligned}$$

Since $\mathbf{q}_i$ remains forever equal to $\mathbf{q}$, $\lim_{i \to \infty} \mathbf{q}_i = \mathbf{q}$. This result brings us to the following definition.

2.5.4 DEFINITION

A Markov probability space $(M, \mathbf{q})$ is said to be *stationary* if $\mathbf{q}_i = \mathbf{q}$ for all i.

Obviously, $(M, \mathbf{q})$ is stationary if and only if $\mathbf{q}$ is an "eigenvector" for M (in the jargon of linear algebra). For those familiar with eigenvectors we state the following proposition, omitting the proof here (cf. Problem 2.5.7).

2.5.5 PROPOSITION

Every transition matrix M has at least one eigenvector, that is, there exists a $\mathbf{q}$ such that $\mathbf{q}M = \mathbf{q}$.

Example 5 Given $(M, \mathbf{q})$ with

$$M = \begin{bmatrix} 1 - \alpha & \alpha \\ \beta & 1 - \beta \end{bmatrix}$$

determine $\mathbf{q}$ so as to make $(M, \mathbf{q})$ stationary.

Solution We need only solve the equation

$$\mathbf{q}M = \mathbf{q}$$

for $\mathbf{q}$. Rewriting this equation as

$$\mathbf{q}(M - I) = 0$$

where I is the identity matrix, we have

$$(p, q)\begin{bmatrix} -\alpha & \alpha \\ \beta & -\beta \end{bmatrix} = (0, 0)$$

Consequently

$$-\alpha p + \beta q = 0$$
$$\alpha p - \beta q = 0$$

Remembering $p + q = 1$, we obtain

$$\alpha p - \beta(1 - p) = 0$$

Hence

$$p = \frac{\beta}{\alpha + \beta}$$

and

$$q = \frac{\alpha}{\alpha + \beta}$$

2.5.6 PROPOSITION

If $\mathbf{q}_\infty$ is the limiting probability vector for $(M, \mathbf{q}_0)$, then $(M, \mathbf{q}_\infty)$ is stationary.

PROOF

From $\mathbf{q}_\infty = \lim_{i \to \infty} \mathbf{q}_0 M^i = \lim_{i \to \infty} [(\mathbf{q}_0 M^{i-1})M] = (\lim_{i \to \infty} \mathbf{q}_0 M^{i-1})M = \mathbf{q}_\infty M$,
we see that $\mathbf{q}_\infty$ satisfies the equation $\mathbf{q}M = \mathbf{q}$. Consequently $(M, \mathbf{q}_\infty)$ is stationary.

Thus, to find $\mathbf{q}_\infty$ for a given $(M, \mathbf{q}_0)$ we must first of all solve the equation $\mathbf{q}M = \mathbf{q}$ for $\mathbf{q}$. If the equation has several solutions, we must further investigate how to eliminate extraneous ones; but even if the solution is unique, we cannot be sure it is indeed the limiting probability vector $\mathbf{q}_\infty$ unless the existence of $\mathbf{q}_\infty$ has already been ascertained for the given $(M, \mathbf{q})$. For example $(M, \mathbf{q}_0)$ in Example 4 of this section does not have $\mathbf{q}_\infty$, but we could all the same solve the equation $\mathbf{q}M = \mathbf{q}$ to find $\mathbf{q} = (\frac{1}{2}, \frac{1}{2})$, which of course is not $\mathbf{q}_\infty$ for the given $(M, \mathbf{q})$.

Given $(M, \mathbf{q}_0)$, if the equation $\mathbf{q}M = \mathbf{q}$ has many solutions, then for each such solution $\mathbf{q}$, $(M, \mathbf{q})$ is stationary with $\mathbf{q}$ being the initial as well as the limiting probability vector. Therefore, if $\mathbf{q}_\infty$ does exist for the given $(M, \mathbf{q}_0)$, $\mathbf{q}_\infty$ does not exist independently of $\mathbf{q}_0$. This gives us a sort of negative result related to question (b).

Our attempts so far to answer questions (a), (b), (c) have produced some interesting results, but they have also revealed the limitations in the purely algebraic approach. On the other hand, other approaches, either geometric or analytic, have turned out to be quite elaborate though their results are more conclusive. We shall therefore state these results without proofs. But first a definition will clarify the issue.

2.5.5 DEFINITION

A Markov probability space $(M, \mathbf{q}_0)$ is said to be *stable* if for all possible initial probability vectors $\mathbf{q}$ (including $\mathbf{q}_0$ in particular) the Markov probability spaces $(M, \mathbf{q})$ have one common limiting probability vector.

In other words stable Markov probability spaces are precisely those which will give an affirmative answer to question (b). The following result, essentially due to Markov, gives a criterion to determine just when a Markov probability space is stable. (A geometric proof of this result can be found in "A Geometric Proof of Markov's Ergodic Theorem," *Proceedings of the American Mathematical Society*, October 1970.)

2.5.1 THEOREM

A Markov probability space $(M, \mathbf{q})$ is stable if and only if for some positive integer n, M^n contains exactly one submatrix which is a transition matrix consisting entirely of positive entries. If M^n itself consists entirely of positive entries, then the limiting probability vector consists entirely of positive components. In this particular case $(M, \mathbf{q})$ is said to be *ergodically stable*.

We conclude this section by applying Markov probability spaces to solving some of the so-called random-walk problems. In particular we shall solve the well-known gambler's ruin problem, which incidentally provides an interesting example of a nonstable Markov probability space.

Example 6 Starting from the ground a man commences his "random walk" up a very tall ladder according to the results of coin tossing. Each time the coin falls heads he moves one step higher. If the coin falls tails he makes no move. Where will the man be after n tosses of the coin? Assume that the coin falls heads with probability p.

Solution Since the man can be anywhere from $K = 0, 1, \ldots$ to $K = n$, we shall use an $(n + 1)$-state Markov probability space with the transition matrix

$$
M = \begin{bmatrix}
q & p & 0 & & \cdots & 0 \\
0 & q & p & 0 & \cdots & 0 \\
0 & 0 & q & p & 0 & \cdots & 0 \\
\cdot & \cdot & \cdot & \cdot & \cdots & \cdot
\end{bmatrix}
$$

and the initial probability vector $\mathbf{q} = (1, 0, 0, \ldots, 0)$ as our model in order to determine the nth probability vector $\mathbf{q}_n$, which will tell us what position the man will be with what probability. Since $\mathbf{q}_n = \mathbf{q}M^n$ and $\mathbf{q} = (1, 0, 0, \ldots)$, $\mathbf{q}_n$ will be just the first row of M^n. Now a few routine multiplications of M by itself reveal that the first row of M^n consists of the familiar $n + 1$ terms in the binomial expansion of $(q + p)^n$, hence

$$
\mathbf{q}_n = (C_0^n q^n p^0, C_1^n q^{n-1} p, \ldots, C_n^n q^0 p^n)
$$

In Section 2.6 we shall refer to $\mathbf{q}_n$ above as the *binomial probability vector* in contrast to the *Poisson probability vector* to be introduced there.

Example 7 (Gambler's Ruin Problem) A and B want to gamble until one of them goes broke. A fair coin is tossed: if it falls heads A takes a dollar from B, if it falls tails B takes a dollar from A. Assuming that initially A and B have a and b dollars, respectively, determine the probability of A going broke before B does.

Solution A and B have between them $K = a + b$ dollars. We shall simulate their joint state of fortune by a particle moving up and down $K + 1$ vertical positions, $0, 1, 2, \ldots, K$, in such a way that the more A takes from B the higher it goes. The particle commences its random walk from the position a, possibly moving up all way to position K to represent gambler B's ruin or down to position 0 to represent gambler A's ruin. When the particle reaches positions 0 or K, it will stay there forever. But conceivably the particle will continue its walk indefinitely. Using an appropriate $(K + 1)$-state Markov probability space with the initial probability vector $\mathbf{q} = (0, 0, \ldots, 1, \ldots, 0)$ where 1 occurs at the ath component, we see that the 0th component of each probability vector $\mathbf{q}_n$ represents the probability that A goes broke within the first n tosses of the coin. These components are clearly nondecreasing and approach a limit as n approaches ∞. This limit, which we can read in the 0th component of the limiting probability vector $\mathbf{q}_\infty$, is the probability we are looking for.

Now the determination of $\mathbf{q}_\infty$ for a given Markov probability space $(M, \mathbf{q})$ cannot be done by blind calculations. We must deduce $\mathbf{q}_\infty$ somehow in finite steps. However, before we get ourselves involved, let us state and interpret the final result:

$$\mathbf{q}_\infty = \left[\frac{b}{a + b}, 0, 0, \ldots, 0, \frac{a}{a + b}\right]$$

which clearly depends very much on the initial

$$\mathbf{q} = (0, 0, \ldots, 1, 0, 0, \ldots, 0)$$

with 1 occurring at the ath position. In other words, assuming that A and B have between them the fixed K dollars, the smaller the initial fortune of A the larger the probability $(K - a)/K$ of his going broke, which is as it should be. From the point of view of stability our Markov probability space $(M, \mathbf{q})$ is a nonstable one; or we might simply say that the $(K + 1) \times (K + 1)$ transition matrix $M = (p_{ij})$ where

$$p_{11} = p_{K+1, K+1} = 1$$
$$p_{i, i+1} = p_{i, i-1} = \tfrac{1}{2} \quad \text{for} \quad i = 2, 3, \ldots, K$$

is nonstable.

Determination of $\mathbf{q}_\infty$

Let $u(a)$ and $v(a)$ be the 0th and the Kth components of $\mathbf{q}_\infty$. We shall determine $u(a)$ first and then deduce $v(a)$ by symmetry. Now trivially

$$u(0) = 1$$
$$u(K) = 0$$

that is, A goes broke with probability 1 if he is broke to begin with, and so on. Since $u(0)$ and $u(K)$ are known, we may be able to find $u(a)$ by somehow relating to them. Accordingly, let us consider

$$u(j) = p(E_1^0 \cup E_2^0 \cup \cdots | E_0^j) \qquad \text{for} \quad j = 1, 2, \ldots, K-1$$

where E_i^0 is the event that A is broke by the ith toss of coin given that his initial fortune is j dollars. By the conditional partition formula (Proposition 2.2.4) we have

$$u(j) = \sum_{k=0}^{K} P(E_1^k | E_0^j) P(E_1^0 \cup E_2^0 \cup \cdots | E_0^j E_1^k)$$

but $P(E_1^k | E_0^j) = \frac{1}{2}$ for $k = j+1$ and $j-1$, and 0 for all other values of k; also

$$P(E_1^0 \cup E_2^0 \cup \cdots | E_0^j E_1^k) = P(E_1^0 \cup E_2^0 \cup \cdots | E_1^k)$$

hence

$$u(j) = \tfrac{1}{2}[u(j+1) + u(j-1)]$$

or

$$u(j+1) - u(j) = u(j) - u(j-1)$$

Repeated application of this formula yields

$$u(j+1) - u(j) = u(1) - u(0)$$

or

$$u(j+1) = u(j) + [u(1) - u(0)]$$

Again, repeated application of this last formula yields

$$u(j+1) = u(1) + j[u(1) - u(0)]$$

Since we already know that $u(0) = 1$, it remains to find $u(1)$. For this let $j = K - 1$ to take advantage of $u(K) = 0$. Thus,

$$u(K) = u(1) + (K-1)[u(1) - u(0)] = 0$$

Consequently,

$$u(1) = \frac{K-1}{K}$$

Now therefore

$$u(j+1) = \frac{K-1}{K} + j\left[\frac{K-1}{K} - 1\right]$$

$$= 1 - \frac{j+1}{K}$$

In other words,

$$u(j) = 1 - \frac{j}{K} \quad \text{for} \quad j = 1, 2, 3, \ldots, K$$

and in particular for $j = a$, we have, remembering that $K = a + b$,

$$u(a) = \frac{b}{a + b}$$

By symmetry we have also

$$v(a) = \frac{a}{a + b}$$

Consequently, since $u(a) + v(a) = 1$, we have

$$\mathbf{q}_\infty = \left[\frac{b}{a + b}, 0, 0, \ldots, 0, \frac{a}{a + b} \right]$$

2.5 PROBLEMS

1. Each year 20% of those driving American cars switch to foreign cars and 10% of those driving foreign cars switch to American cars. If 60% of all motorists are driving American cars this year, what percentage of the motorists will be driving American cars 5 years from now?

2. A signal of either A or B is transmitted through a series of relay stations. While each relay station makes no error in transmitting exactly the signal received, it receives a wrong signal with probability 0.1. If the signal sent out by the originating station is A, what is the probability of the third relay station receiving the signal A? Of the first three relay stations all receiving the signal A?

3. Show that a two-state Markov probability space $(M, \mathbf{q})$ is a Bernoulli probability space if both rows of M are $\mathbf{q}$. Hint: Show that the conditions of Definition 2.4.1 are satisfied.

4. Given a Markov probability space $(M, \mathbf{q})$ with

$$M = \begin{bmatrix} \frac{1}{2} & \frac{1}{2} & 0 \\ \frac{1}{2} & \frac{1}{2} & 0 \\ 0 & \frac{1}{2} & \frac{1}{2} \end{bmatrix} \quad \text{and} \quad \mathbf{q} = (0, 0, 1)$$

calculate the conditional probability $P(E_1^3 | E_3^3)$.

5. Given a Markov probability space $(M, \mathbf{q})$ with

$$M = \begin{bmatrix} \frac{1}{4} & \frac{1}{4} & \frac{1}{2} \\ 0 & \frac{2}{3} & \frac{1}{3} \\ \frac{3}{4} & 0 & \frac{1}{4} \end{bmatrix}$$

find $\mathbf{q}$ so that $(M, \mathbf{q})$ is stationary.

6. A man commences a random walk from position 1 to positions 2, 3, ... From position j he moves to position $j + 1$ with probability $1/(j + 1)$ or else stays at position j for $j = 1, 2, 3, \ldots$. Show that after n units of transition time the probability of his remaining in the first N positions is less than $N/2^n$. What happens to this probability when n tends to ∞?

7. Optional. Given a $k \times k$ transition matrix M show that there exists a k-dimensional nonzero row vector $\mathbf{v}$ such that $\mathbf{v}M = \mathbf{v}$. Hint: Let $M = (p_{ij})$ and $\mathbf{v} = (v_1, v_2, \ldots, v_k)$. Write out the k simultaneous linear equations by expanding $\mathbf{q}M = \mathbf{q}$ and show that the k equations are not independent by adding the left-hand sides and the right-hand sides of the k equations separately. The resulting identity shows that the k linear equations are not independent, hence there exists at least one nontrivial solution $\mathbf{v}$.

8. A square matrix with nonnegative entries is called a *stochastic matrix* (or transition matrix) if all the row sums are equal to 1. Show that if M is a stochastic matrix so is M^n for $n = 1, 2, 3, \ldots$.

9. If $(M, \mathbf{q})$ is a stable Markov probability space, then the stochastic matrix M is called *stable*.
   a. Show that if M is a stable stochastic matrix, then $\lim_{n \to \infty} M^n = M^\infty$ is a stochastic matrix consisting of identical rows $\mathbf{q}_\infty$.
   b. Conversely show that if a stochastic matrix M is such that $\lim_{n \to \infty} M^n$ is a matrix consisting of identical rows, then $(M, \mathbf{q})$ is a stable Markov probability space. We thus obtain a nonprobabilistic characterization of a stable stochastic matrix as a stochastic matrix whose infinite power consists of identical rows.

10.a. Use Markov's theorem (Theorem 2.5.1) to show that the stochastic matrix

$$M = \begin{bmatrix} 0 & \alpha & \beta \\ \alpha & \beta & 0 \\ \alpha & \beta & 0 \end{bmatrix}$$

where $\alpha + \beta = 1$ and $\alpha\beta > 0$, is stable.
   b. Determine M^∞.

2.6 POISSON PROBABILITY SPACES

In constructing a probability space $(S, \mathscr{S}, P)$ the sample set S is easily decided upon according to the observed natures of outcomes of the given random phenomenon. If we observe each outcome with a *discrete* timing $t = 1, 2, 3, \ldots$, naturally the sample set should consist of infinite sequences of numbers (or whatever observed values). On the other hand if we observe an outcome *continuously*, then our sample set should consist of functions in a continuous t and taking whatever observed values such as integers or real numbers.

We shall construct a Poisson probability space $(S, \mathscr{S}, P)$ with S the set of nonnegative integer-valued functions $x(t)$ of a nonnegative real variable t. Such a sample set is suitable for modeling a wide variety of "phenomena of random occurrences," for example, incidences of highway accidents, fluctuating lengths of a line of people, rise-and-fall of a population due to births and deaths. We shall assume that $x(t) = 0$ at $t = 0$ reflecting the fact that we start counting the numbers of accidents, the lengths of a line of people, the sizes of a population, and so on, at $t = 0$. The primary events of interest to us are represented by E_t^j consisting of all functions assuming the integer value j at the time t. For example, E_t^j may represent the event that exactly j accidents have been counted up to the specific moment t, or that a line of j people is formed at t. Consistent with the choice of $\mathscr{S}$ for Bernoulli and Markov probability spaces, we shall again let $\mathscr{S}$ be the sigma algebra of subsets of $\mathscr{S}$ generated by these primary events E_t^j ; thus in particular the intersection events $E_{t_1}^{j_1} E_{t_2}^{j_2} \cdots E_{t_n}^{j_n}$ for $0 < t_1 < t_2 < \cdots < t_n$ and for non-negative integers $j_1, j_2, \ldots, j_n$ are included in $\mathscr{S}$ and so are any finite or countable union of these intersection events.

This leaves us the construction of P. Again we shall first make a few reasonable assumptions on P reflecting the nature of the type of random phenomena under observation. Naturally, different types of random phenomena will give rise to different assumptions on P leading thus to different probability measures P and hence to different types of probability spaces $(S, \mathscr{S}, P)$. The Poisson probability spaces are the result of adopting a highly restrictive set of assumptions on P, which we shall presently consider in detail. In developing these assumptions we should always keep in mind the type of random phenomena we are trying to describe. These are, for example, the cumulative incidence of accidents, the cumulative number of cars arriving at a toll-collecting station, and the total number of α particles emitted by some radioactive substance. Other examples mentioned earlier, such as the fluctuating length of a line and the rise and fall of a population, will require less restrictive assumptions on P, and consequently will not be modeled by Poisson probability spaces.

What assumptions can we make on P without seriously distorting our image of the random phenomenon under observation? These are:

(1) $P(E_0^0) = 1$, since we count everything from $t = 0$ ignoring whatever may have happened before $t = 0$.

(2) $P(E_{s+t}^{j+k} \mid D E_s^j) = \alpha_k(t)$ and $\sum_{k=0}^{\infty} \alpha_k(t) = 1$ for each $t > 0$, where D is any member of the sigma algebra generated by the "earlier" primary events E_r^i for $r < s$ and $i = 0, 1, 2, \ldots$. This statement may be interpreted as saying that the probability of k incidents occurring during the time interval $[s, s + t]$ depends only on the length t of this interval and not on how many incidents there had been or how these incidents had occurred prior to the moment s, since the expression $\alpha_k(t)$ leaves out s and j completely. Obviously, such an assumption would be quite out of order if we were counting the size of a population, since the larger the population at the moment s the larger the number of births during a time interval immediately following s. The assumption

$$\sum_{k=0}^{\infty} \alpha_k(t) = 1$$

which automatically excludes any positive values for $P(E_{s+t}^{j+k} | E_s^j)$ for negative k, reflects the fact that the cumulative number of incidents cannot decrease. Again such an assumption would have been quite out of question for, say, the fluctuating lengths of a line of people waiting at a service counter.

We shall now make some further assumptions on the *transition probability function* $\alpha_k(t)$ in order to facilitate our ultimate determination of P. Some of these assumptions are of purely mathematical nature (e.g., continuity and differentiability), the justification of which comes mainly from the adequacy of the final $(S, \mathscr{S}, P)$ as a mathematical model describing the random phenomenon in question.

(3) $\lim_{t \to 0} \alpha_k(t)$ exists for each $k = 0, 1, 2, \ldots$. Taking these limits as the values of hitherto undefined $\alpha_k(0)$, we further postulate

$$\alpha_k(0) = \begin{cases} 1 & \text{for} \quad k = 0 \\ 0 & \text{for} \quad k \geq 1 \end{cases}$$

There is nothing mysterious in the above assumption $\lim_{t \to 0} \alpha_0(t) = 1$ since as the time interval $[s, s + t]$ shrinks toward s we may expect the probability of no increase in incidents to approach 1.

(4) $\alpha_k(t)$ is differentiable at $t = 0$ for each $k = 0, 1, 2, \ldots$. Denoting these derivatives by $\alpha_k'(0)$, we further postulate

$$\alpha_k'(0) = \begin{cases} \lambda & \text{for} \quad k = 1 \\ 0 & \text{for} \quad k \geq 2 \end{cases}$$

where λ is some positive constant. Again there is nothing mysterious in the above assumption since for a very small interval $[s, s + t]$ we may assume that the probability of more than 1 incident is negligible and that the probability of exactly 1 incident is just about proportional to the length of the interval t. Note that $\alpha_1'(0) = \lambda$ is equivalent to

$$\alpha_1(t) = \lambda t + \varepsilon t$$

where ε approaches 0 as t approaches 0.

As a consequence of (4), which did not say much about $\alpha_0'(0)$, we have

(4a) $\alpha_0'(0) = -\lambda$. This follows immediately from the fact that $\sum_{k=0}^{\infty} \alpha_k'(0) = 0$ [consequence of (2)] and (4). It says essentially that $\alpha_0(t)$ is monotone decreasing at $t = 0$, that is, the probability of no incidents in the interval $[s, s + t]$ decreases as t increases. Henceforth condition (4a) will be regarded as part of assumption (4).

With the preceding four assumptions on P we now proceed to determine the values of P on (i) the primary events E_t^j; (ii) the events F_t where each F_t is a countable union of E_t^j with t fixed; (iii) the finite intersection events of events F_t; and (iv) all events in the sigma algebra $\mathscr{S}$ generated by the primary events.

The first step toward determining $P(E_t^k)$ is to note

(5) $P(E_t^k) = P(E_{0+t}^{0+k} | E_0^0) = \alpha_k(t)$

with conditions (1) and (2) respectively justifying the first and the second equalities. The next step is to determine $\alpha_0(t)$ and then relate all $\alpha_k(t)$ ultimately to $\alpha_0(t)$. We shall first find a recurrence formula relating $\alpha_k(t)$ to $\alpha_{k-1}(t)$ via statements (6) and (7) to follow:

$$(6) \qquad \alpha_k(t + \Delta t) = \sum_{h=0}^{k} \alpha_h(t)\alpha_{k-h}(\Delta t)$$

To prove (6) we need only use the partition formula (Proposition 2.2.2),

$$\alpha_k(t + \Delta t) = P(E_{t+\Delta t}^k)$$

$$= \sum_{h=0}^{k} P(E_t^h)P(E_{t+\Delta t}^k \mid E_t^h)$$

$$= \sum_{h=0}^{k} \alpha_h(t)\alpha_{k-h}(\Delta t)$$

in view of (5) and (2).

$$(7) \qquad \alpha_k'(t) = \lambda\alpha_{k-1}(t) - \lambda\alpha_k(t) \qquad \text{for} \quad k \geq 1$$

To see this, consider

$$\alpha_k'(t) = \lim_{\Delta t \to 0} \frac{\alpha_k(t + \Delta t) - \alpha_k(t)}{\Delta t}$$

$$= \lim_{\Delta t \to 0} \frac{\sum_{h=0}^{k} \alpha_h(t)\alpha_{k-h}(\Delta t) - \alpha_k(t)}{\Delta t}$$

$$= \lim_{\Delta t \to 0} \left[\frac{\sum_{h=0}^{k-1} \alpha_h(t)\alpha_{k-h}(\Delta t)}{\Delta t} + \frac{\alpha_k(t)\alpha_0(\Delta t) - \alpha_k(t)}{\Delta t} \right]$$

$$= \sum_{h=0}^{k-1} \alpha_h(t)\alpha_{k-h}'(0) + \alpha_k(t)\alpha_0'(0)$$

$$= \lambda\alpha_{k-1}(t) - \lambda\alpha_k(t)$$

where (6) is used for the second equality and (4) is used for the last equality. Equation (7) now leads to the recurrence formula

$$(8) \qquad \alpha_k(t) = e^{-\lambda t} \int_0^t \lambda e^{\lambda t} \alpha_{k-1}(t) \, dt \qquad \text{for} \quad k \geq 1$$

To see this, multiply (7) by $e^{\lambda t}$ to obtain

$$e^{\lambda t}\alpha_k'(t) + \lambda e^{\lambda t}\alpha_k(t) = \lambda e^{\lambda t}\alpha_{k-1}(t)$$

or

$$\frac{d}{dt}[e^{\lambda t}\alpha_k(t)] = \lambda e^{\lambda t}\alpha_{k-1}(t)$$

hence

$$e^{\lambda t}\alpha_k(t) = \int_0^t \lambda e^{\lambda t}\alpha_{k-1}(t)\, dt + c_k$$

or

$$\alpha_k(t) = e^{-\lambda t}\int_0^t \lambda e^{\lambda t}\alpha_{k-1}(t)\, dt + c_k e^{-\lambda t}$$

The constant c_k turned out to be 0 as we set $t = 0$, $k = 1, 2, 3, \ldots$.

At last we are ready for the following proposition.

2.6.1 PROPOSITION

In the construction of a Poisson probability space $(S, \mathscr{S}, P)$, if P satisfies conditions (1) to (4), then

$$P(E_t^k) = e^{-\lambda t}\frac{(\lambda t)^k}{k!} \quad \text{for} \quad k = 0, 1, 2, \ldots$$

PROOF

First, in view of (5), we determine

$$P(E_t^0) = \alpha_0(t)$$

But by (6) we have

$$\alpha_0(t + \Delta t) = \alpha_0(t)\alpha_0(\Delta t)$$

consequently

$$
\begin{aligned}
\alpha_0'(t) &= \lim_{\Delta t \to 0} \frac{\alpha_0(t + \Delta t) - \alpha_0(t)}{\Delta t} \\
&= \lim_{\Delta t \to 0} \frac{\alpha_0(t)[\alpha_0(\Delta t) - 1]}{\Delta t} \\
&= \alpha_0(t)\alpha_0'(0) \\
&= -\lambda\alpha_0(t)
\end{aligned}
$$

where (3) $\alpha_0(0) = 1$ is used for the third equality and (4a) $\alpha_0'(0) = -\lambda$ is used for the last equality. Now since

$$\frac{d\alpha_0}{dt} = -\lambda\alpha_0$$

α_0 must be of the form

$$\alpha_0(t) = ce^{-\lambda t}$$

in which c must be 1 since $\alpha_0(0) = 1$.

Having determined $\alpha_0(t) = e^{-\lambda t}$, we next determine $\alpha_1(t), \alpha_2(t), \ldots$ by successive applications of recurrence formula (8). Indeed,

$$\alpha_1(t) = e^{-\lambda t} \int_0^t \lambda e^{\lambda t} e^{-\lambda t}\, dt$$
$$= e^{-\lambda t}(\lambda t)$$

and in general if

$$\alpha_k(t) = e^{-\lambda t} \frac{(\lambda t)^k}{k!}$$

then

$$\alpha_{k+1}(t) = e^{-\lambda t} \int_0^t \lambda e^{\lambda t} e^{-\lambda t} \frac{(\lambda t)^k}{k!}\, dt$$
$$= e^{-\lambda t} \frac{(\lambda t)^{k+1}}{(k+1)!}$$

and this completes the proof.

2.6.2 PROPOSITION

In the construction of a Poisson probability space $(S, \mathscr{S}, P)$ if P satisfies conditions (1) to (4), then

$$P(E_{t_1}^{j_1} E_{t_2}^{j_2} \cdots E_{t_m}^{j_m}) = e^{-\lambda t_m} \lambda^{j_m} \prod_{i=1}^m \frac{(t_i - t_{i-1})^{j_i - j_{i-1}}}{(j_i - j_{i-1})!}$$

where $0 < t_1 < t_2 < \cdots < t_m$.

The only important thing about this proposition is that the probability of an intersection event of a finite number of primary events can be unequivocally determined; the precise expression of this probability need not concern us.

PROOF

By the extended product formula (Proposition 2.2.1) and in view of (2), we have

$$P(E_{t_1}^{j_1} E_{t_2}^{j_2} \cdots E_{t_m}^{j_m}) = P(E_{t_1}^{j_1}) P(E_{t_2}^{j_2} \mid E_{t_1}^{j_1}) \cdots P(E_{t_m}^{j_m} \mid E_{t_{m-1}}^{j_{m-1}})$$
$$= \alpha_{j_1}(t_1) \alpha_{j_2 - j_1}(t_2 - t_1) \cdots \alpha_{j_m - j_{m-1}}(t_m - t_{m-1})$$

After substituting $e^{-\lambda t_1}(\lambda t_1)^{j_1}/j_1!$, and so on, for $\alpha_{j_1}(t_1)$, and so on, we see that the rest of the computation amounts to a simple exercise.

Looking back at our programs (i) to (iv) for the construction of P (cf. page 68), we see that Proposition 2.6.1 takes care of (i) and by a simple " additive extension " also (ii) and that Proposition 2.6.2 essentially takes care of (iii) since any finite intersection event of events of type (ii) can be expressed as a countable union of disjoint events considered in Proposition 2.6.2. Finally we can check that events of type (iii) constitute a Boolean algebra (hence easily a semialgebra) on which P is additively defined (i.e., not in violation of the axioms of measure), so that by the extension principle of measure (page 47) P is defined on the sigma algebra $\mathscr{S}$ generated by the primary events. We shall not burden ourselves with the full details leading to this last step (iv).

Our informal construction of $(S, \mathscr{S}, P)$ substantiates the following definition.

2.6.1 DEFINITION

By a Poisson probability space we mean a probability space $(S, \mathscr{S}, P)$ in which S consists of all nonnegative integer-valued functions $x(t)$ of a nonnegative real variable t; $\mathscr{S}$ is the sigma algebra of subsets of S generated by the subsets

$$E_t^j = \{x(t): x(t) = j\}$$

for $t \geq 0$ and $j = 0, 1, 2, \ldots$; and P is a probability measure satisfying the following conditions:

(1) $P(E_0^0) = 1$
(2) $P(E_{s+t}^{j+k} \mid DE_s^j) = \alpha_k(t)$

where D is any member of the sigma algebra generated by all E_r^i for $r < s$ and $i = 0, 1, 2, \ldots$, and the $\alpha_k(t)$ are defined for $t \geq 0$ such that
(a) $\sum_{k=0}^{\infty} \alpha_k(t) = 1$ for each $t > 0$;
(b) $\alpha_k(t)$ are all continuous at $t = 0$ and

$$\alpha_k(0) = \begin{cases} 1 & \text{for} \quad k = 0 \\ 0 & \text{for} \quad k \geq 1 \end{cases}$$

(c) $\alpha_k(t)$ are all differentiable at $t = 0$ and

$$\alpha_k'(0) = \begin{cases} \lambda & \text{for} \quad k = 1 \\ 0 & \text{for} \quad k \geq 2 \end{cases}$$

where λ is a positive constant.

Like any respectable axiomatic definition, the definition of Poisson probability space keeps the postulates (or axioms) at the bare minimum. Everything else we have said about the Poisson probability space including Propositions 2.6.1–2 can be derived from the postulates in the definition.

In adopting a Poisson probability space as a mathematical model for dealing with a given random phenomenon we must first decide on a specific value of λ. Since λ is the only unspecified item in Definition 2.6.1 once this value is agreed upon a unique Poisson probability space will come into focus. With the choice of λ the transition from the real world to the world of mathematical model is accomplished in one stroke. The all-important choice of λ is made possible by the following concrete interpretation of λ. Recall from page 68 that

$$\alpha_1(t) = \lambda t + \varepsilon t$$

That is, for a very small value of t we have

$$\lambda \doteq \frac{\alpha_1(t)}{t}$$

Now $\alpha_1(t)$ is the probability of exactly 1 incident occurring in the interval of length t. We can actually estimate this probability as follows. Let us suppose that over the time interval of length T an average of N incidents occur (N can be estimated empirically). If we divide this time interval into n sufficiently small subintervals of length $t = T/n$, in each subinterval at most 1 incident will be observed; thus 1 incident is observed in N out of n subintervals and we may therefore estimate

$$\alpha_1(t) = \frac{N}{n}$$

Consequently we may estimate

$$\lambda = \frac{N}{nt} = \frac{N}{T}$$

In other words λ may be interpreted as the average number of incidents per unit time since N was the average number of incidents over T units of time.

Example 1 A toll collector knows from years of experience that on the average 30 cars pass the toll station during the noon hour. Use this information to calculate the probability that no cars will pass the station between 12:30 and 12:34 p.m.

Solution Since on the average $\frac{30}{60} = 0.5$ cars per minute pass by, we take $\lambda = 0.5$ for our Poisson probability space, in which t represents minutes and $x(t)$ represents the number of cars passed by during t minutes after 12:30. In terms of our model we need only calculate $\alpha_0(4)$. By Proposition 2.6.1 we have

$$\alpha_0(4) = e^{-(0.5)(4)} = e^{-2} \doteq 0.14$$

Example 2 A typist makes about a dozen mistakes per hour. If it takes 5 minutes to type a page, what is the probability of exactly one mistake occurring in a typed page?

Solution We use a Poisson probability space in which $x(t)$ represents the number of mistakes occurring in t minutes. Letting $\lambda = \frac{12}{60} = 0.2$, we see that

$$\alpha_1(5) = e^{-(0.2)(5)} = e^{-1} \doteq 0.37$$

by Proposition 2.6.1.

We conclude this section with a joint consideration of Poisson probability vector and the binomial probability vector mentioned in the preceding section. This will lead us in a natural way to the well-known Poisson approximation theorem.

2.6.2 DEFINITION

By a *Poisson probability vector* we mean a countably infinite dimensional row vector

$$\mathbf{q} = (q_0, q_1, q_2, \ldots)$$

such that

$$q_k = e^{-\mu} \frac{\mu^k}{k!}$$

for $k = 0, 1, 2, \ldots$ where μ is some positive constant.

By a *binomial probability vector* we mean a countably infinite dimensional row vector

$$\mathbf{q} = (q_0, q_1, q_2, \ldots, q_n, \ldots)$$

such that

$$q_k = \begin{cases} C_k^n p^k (1-p)^{n-k} & \text{for} \quad k = 0, 1, \ldots, n \\ 0 & \text{for} \quad k \geq n + 1 \end{cases}$$

where p is some positive constant between 0 and 1, and n is a positive integer.

Clearly if $(S, \mathscr{S}, P)$ is a Poisson probability space, then for any $t > 0$

$$[P(E_t^0), P(E_t^1), P(E_t^2), \ldots]$$

forms a Poisson probability vector since

$$P(E_t^k) = \frac{e^{-\lambda t}(\lambda t)^k}{k!}$$

and we need only let $\lambda t = \mu$.

Likewise, if $(S, \mathscr{S}, P)$ is a Markov probability space describing the random walk considered in Section 2.5, Example 6, then for any n

$$[P(E_n^0), P(E_n^1), \ldots, P(E_n^n), 0, 0, \ldots]$$

forms a binomial probability vector since

$$P(E_n^k) = C_k^n p^k (1-p)^{n-k}$$

The fact that the Poisson probability vector and the binomial probability vector may be closely related can be seen from the fact that each Poisson probability space carries within it a Markov probability space which resembles the one that gave rise to the binomial probability vector. We recall that each element in the sample set of a Poisson probability space is a function $x(t)$; by restricting t to integer values 0, 1, 2, ..., we obtain an infinite sequence $x(1)$, $x(2)$, $x(3)$, ..., the kind of element considered in the sample set of a Markov probability space. Now within the Poisson probability space we have

$$P(E_{i+1}^{j+k} \mid E_i^j) = \alpha_k(1) = \begin{cases} 1 - \lambda + \varepsilon_0 & \text{for} \quad k = 0 \\ \lambda + \varepsilon_1 & \text{for} \quad k = 1 \\ \varepsilon_2 & \text{for} \quad k \geq 2 \end{cases}$$

where ε_0, ε_1, ε_2 are relatively small compared to λ. This is essentially a paraphrase of conditions (4) and (4a). On the other hand, within the Markov probability space describing the random walk we have

$$P(E_{i+1}^{j+k} \mid E_i^j) = \begin{cases} 1 - p & \text{for} \quad k = 0 \\ p & \text{for} \quad k = 1 \\ 0 & \text{for} \quad k \geq 2 \end{cases}$$

If we identify p with λ above and ignore ε_0, ε_1, ε_2, the resemblance between the two sets of transition probabilities is unmistakable. As a consequence of this we expect the two corresponding probability vectors at $t = n$ to resemble one another. In other words, $\mathbf{q} = (q_0, q_1, \ldots)$ with

$$q_k = \frac{e^{-\lambda n}(\lambda n)^k}{k!}$$

must resemble $\mathbf{q} = (q_0, q_1, \ldots)$ with

$$q_k = C_k^n p^k (1 - p)^{n-k}$$

provided that $p = \lambda$. We expect therefore to have

$$C_k^n p^k (1 - p)^{n-k} \doteq \frac{e^{-np}(np)^k}{k!}$$

or

$$e^{-\mu} \frac{\mu^k}{k!} \doteq C_k^n \left(\frac{\mu}{n}\right) \left(1 - \frac{\mu}{n}\right)^{n-k}$$

In order for the above approximation to be close, the three ε's appearing in the transition probabilities of the Poisson probability space have to be small. In fact the only way to improve the approximations is to force the ε's to tend to 0. We can do this by choosing the unit of time t to be increasingly smaller (second, millisecond, microsecond, etc.) When this is done, λ cannot but go to 0 also. Thus we may expect the first of the above approximations to be good only when $p = \lambda$ is very small, and the second approximation to be good only when n is very large, and this latter is precisely what is asserted by the following theorem.

2.6.1 THEOREM (Poisson Approximation Theorem)

For any $\mu > 0$ and for each $k = 0, 1, 2, \ldots, n$, we have

$$\lim_{n \to \infty} C_k^n \left(\frac{\mu}{n}\right)^k \left(1 - \frac{\mu}{n}\right)^{n-k} = \frac{e^{-\mu}\mu^k}{k!}$$

PROOF
We have after rearrangement of factors

$$C_k^n \left(\frac{\mu}{n}\right)^k \left(1 - \frac{\mu}{n}\right)^{n-k} = \left(1 - \frac{\mu}{n}\right)^{n-k} \mu^k \frac{1}{k!} \frac{n(n-1) \cdots (n-k+1)}{n^k}$$

but

$$\lim_{n \to \infty} \left(1 - \frac{\mu}{n}\right)^{n-k} = e^{-\mu} \quad \text{and} \quad \lim_{n \to \infty} \frac{n(n-1) \cdots (n-k+1)}{n \cdot n \cdots \cdots n} = 1$$

and the theorem is proved.

Although the Poisson approximation theorem can thus be proved precisely, this theorem obscures the fact that what makes the approximation good is not so much the largeness of n as the smallness of $p = \mu/n$. In general, it is known that the approximation is good as long as $p < 0.1$; actually this is true even when n is not large. For instance, for $n = 2$, $\mu = 0.02$, and $p = 0.01$, we have the following rather remarkable approximations:

$$C_0^2 (0.01)^0 (1 - 0.01)^2 = 0.9801 \doteq 0.9802 = \frac{e^{-0.02}(0.02)^0}{0!}$$

$$C_1^2 (0.01)^1 (1 - 0.01)^1 = 0.0198 \doteq 0.0196 = \frac{e^{-0.02}(0.02)^1}{1!}$$

$$C_2^2 (0.01)^2 (1 - 0.01)^0 = 0.0001 \doteq 0.0002 = \frac{e^{-0.02}(0.02)^2}{2!}$$

Example 3 From a 100-page book known to contain 5 misprints, a leaf (2 pages) is selected at random. What is the probability that no misprints are found in it?

Solution The probability of each misprint occurring elsewhere from the leaf is $p = 98/100$. Therefore the probability of all 5 misprints occurring elsewhere from the leaf is $(98/100)^5$ or by Poisson approximation

$$C_5^5 \left(\frac{98}{100}\right)^5 \left(\frac{2}{100}\right)^0 = C_0^5 \left(\frac{2}{100}\right)^0 \left(\frac{98}{100}\right)^5$$

$$\doteq \frac{e^{-0.1}(0.1)^0}{0!} = 0.90$$

Alternatively, if we use a Poisson probability space with $\lambda = 5/100$, then $\alpha_0(2) = e^{-(5/100)\cdot 2} = e^{-0.1}$ and we obtain the same result.

2.6 PROBLEMS

1. From a 500-page book known to contain 25 misprints, 3 pages are selected at random. What is the probability of finding 1 or more misprints in them? Use an appropriate Poisson probability space as a model.

2. In a field of 5000 sq. ft, 250 poison mushrooms are found. Calculate the probability of exactly 1 poison mushroom in an arbitrarily chosen lot of 10 sq. ft.

3. From the dough mixed with 1000 raisins, 100 loaves of bread are made. Calculate the probability of at least 10 raisins being in a loaf.

4. An average of 96 typewriters are sold at a department store each year. How many typewriters should be in stock at the beginning of the year in order that the store may be 99% sure of not running out of typewriters during the year?

5. In a supermarket it is estimated that about 75 shoppers appear at the cashier's counter during the noon hour. Assuming that 2 minutes is required to take care of each customer, determine the number of counters that must be open in order to be 95% sure that no customers wait in line. Use Poisson approximations.

6. Do Problem 2.6.5 replacing 75 with 150 shoppers. Is it necessary to open twice as many counters?

7 a. A typist who makes about a dozen mistakes per hour has just made a mistake. What is the probability of her making the next mistake within 1 minute?
    b. The same typist made her last mistake 10 minutes ago. What is the probability of her making the next mistake within 1 minute? Looking at the answers for (a) and (b), do you think that we are being a little unrealistic in adopting the Poisson probability space as our model?

8. Birth probability space. As a slight generalization of Poisson probability space (cf. Definition 2.6.1), consider the following so-called (pure) birth probability space. By a birth probability space we mean a triple $(S, \mathscr{S}, P)$ in which S consists of all non-negative integer-valued function $x(t)$ of a nonnegative (time) variable t; $\mathscr{S}$ is the sigma algebra of subsets of S generated by the subsets

$$E_t^j = \{x(t): x(t) = j\}$$

for $t \geq 0$, and $j = 0, 1, 2, \ldots$; and P is a probability measure satisfying the following conditions:
    (1) $P(E_0^a) = 1$ where a is a fixed nonnegative integer.
    (2) $P(E_{s+t}^{j+k} \mid DE_s^j) = \alpha_{j, k}(t)$ where D is any member of the sigma algebra generated by E_r^i for $r < s$ and $i = 0, 1, 2, \ldots$, and the $\alpha_{j, k}(t)$ are defined for each $t \geq 0$ such that for each $j \geq a$,
    (a) $\sum_{k=0}^{\infty} \alpha_{j, k}(t) = 1$ for each $t \geq 0$.

(b) The $\alpha_{j,k}(t)$ are all continuous at $t = 0$ and

$$\alpha_{j,k}(0) = \begin{cases} 1 & \text{for} \quad k = 0 \\ 0 & \text{for} \quad k \geq 1 \end{cases}$$

(c) The $\alpha_{j,k}(t)$ are all differentiable at $t = 0$ and

$$\alpha'_{j,k}(0) = \begin{cases} \lambda_j & \text{for} \quad k = 1 \\ 0 & \text{for} \quad k \geq 2 \end{cases}$$

where λ_j is a positive constant depending on j. (In general, λ_j increases monotonically on j.)

Now consider the following questions:
  (i) Compare the preceding definition of birth probability space with Definition 2.6.1 for the Poisson probability space. In what way is a birth probability space a generalization of the Poisson probability space?
  (ii) For what sort of random phenomena is a birth probability space an appropriate model?
  (iii) How would you interpret the statement that $\lambda_j = \alpha'_{j,1}(0)$ increases monotonically on j?
  (iv) Show that (c) entails $\alpha'_{j,0}(0) = -\lambda_j$ in view of (a).

  9. Yule probability space. A well-known special case of birth probability space (cf. Problem 2.6.8) is the so-called Yule probability space, in which $\lambda_j = j\lambda$ where λ is a fixed positive constant. For such a probability space, proceed as follows to determine $P(E_t^{a+k})$.
  a. Show that $P(E_t^{a+k}) = \alpha_{a,k}(t)$ where $\alpha_{a,k}(t)$ is as defined in Problem 2.6.8(2).
  b. Establish the differential equation for $\alpha_{a,k}(t)$,

$$\alpha'_{a,k}(t) = (a + k - 1)\lambda\alpha_{a,k-1}(t) - (a + k)\lambda\alpha_{a,k}(t)$$

for $k \geq 1$.
  c. Solve the above differential equation to obtain the recurrence formula

$$\alpha_{a,k}(t) = e^{-(a+k)\lambda t} \int_0^t (a + k - 1)\lambda e^{(a+k)\lambda t} \alpha_{a,k-1}(t)\, dt$$

for $k \geq 1$.
  d. Determine for $k = 0$

$$\alpha_{a,0}(t) = e^{-a\lambda t}$$

  e. Use the recurrence formula in (c) to obtain for $k = 1$ and 2

$$\alpha_{a,1}(t) = ae^{-a\lambda t}(1 - e^{-\lambda t})$$

$$\alpha_{a,2}(t) = \frac{(a+1)a}{2!} e^{-a\lambda t}(1 - e^{-\lambda t})^2$$

f. Show by induction

$$\alpha_{a,\,k}(t) = C_k^{a+k-1} e^{-a\lambda t}(1 - e^{-\lambda t})^k$$

for $k \geqq 0$. Here, $P(E_k^{a+k}) = \alpha_{a,\,k}(t)$ for $k = 0, 1, 2, \ldots$ constitute the (a, λ) *Yule probability vector* (or distribution).

Chapter 3
RANDOM VARIABLES AND PROBABILITY DISTRIBUTIONS

This chapter introduces a new type of probability models, namely the random variables. These models are subsequently generalized to random vectors, and ultimately to a more elaborate model known as random functions, or mathematical stochastic processes. Random variables and random vectors are all built on the basic model of probability spaces; they are essentially mappings of probability spaces into the traditional Euclidean spaces. These mappings have the interesting effect of converting the Euclidean spaces into probability spaces reflecting the random natures of these random variables. Each probability measure induced on the Euclidean space by a given random variable is called the probability distribution of this random variable.

The preceding outline may acquire a more concrete meaning as we see in Section 3.1 (Probability Distribution of a Random Variable) how a random variable actually arises in a natural context. In Section 3.2 (Distribution Functions), we introduce a convenient way of describing the probability distribution of a random variable. Section 3.3 (Transformations of a Random Variable) shows how a random variable may be altered, or how a random variable may be regarded as an alteration of another. If a random variable is altered, so must its probability distribution be; but the new distribution may be easily related to the old distribution. This is shown in Section 3.3.

In Section 3.4 (Random Vector, Family of Random Variables), we consider several random variables simultaneously to arrive at a random vector, which can then be regarded as a mapping of the probability space into a higher-dimensional Euclidean space. After replacing the real line with the plane or a higher-dimensional space, one may investigate a random vector in much the same way as we did a random variable. But, eventually, problems peculiar to a random vector emerge, and we must make preparations for handling such problems. This will be done in Chapter 5.

Section 3.5 (Singular Distribution and Decomposition of Distributions) treats the unusual (or pathological) random variables and vectors that are primarily of academic interest. However, familiarity with these may give one a sense of completeness in his understanding of probability distributions. In Section 3.5 we also consider distribution functions for random vectors. All together, Sections 3.4-3.6 may be regarded as somewhat parallel to Sections 3.1-3.3. Section 3.6 (Borel Functions of a Random Vector) contains some optional material (pages 133–138) that may be omitted in the first reading.

3.1 PROBABILITY DISTRIBUTION OF A RANDOM VARIABLE

In observing outcomes of a random experiment, a certain numerical aspect may be singled out for consideration. For example, in the experiment of repeatedly tossing a coin we may count the number N of heads in each sequence of tosses; in observing the passing of cars at a toll-collecting station we may note the arrival time T of the first passing car; in daily observations of the water level at a pier we may measure the water level H at sunset. The variables N, T, H, and many others all assume random values, and are therefore referred to as random variables. For each random experiment we may consider any number of random variables. For example, in observing the passing of cars we may consider the total number of cars passed, the total amount of toll collected, the time lag between the first and the second cars, and so on. Since each random variable assumes a specific value with each outcome, it is natural to regard a random variable as a real-valued function having the set of all outcomes as its domain.

Given a random variable X, we would naturally want to know what the probability is that X assumes a certain value or what the probability is that X assumes any value in a certain interval. More generally, for each given subset B of the real line we want to know the probability that X assumes any value in B. Now the *event* that X assumes any value in B can be represented by a certain subset of the sample set S, namely (see Fig. 3.1.1)

$$X^{-1}(B) = \{\omega \in S: X(\omega) \in B\}$$

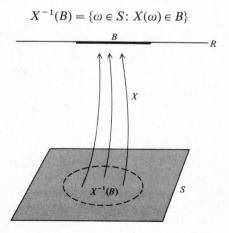

Figure 3.1.1

Therefore if $X^{-1}(B)$ happens to belong to the family $\mathscr{S}$ of considered events, the probability of this event is already defined. On the other hand, if $X^{-1}(B)$ does not belong to $\mathscr{S}$, we may have to extend $\mathscr{S}$ to include $X^{-1}(B)$. However, much complication can be avoided if we refrain from choosing too impractical a set B. Technically we shall confine ourselves to Borel sets B. Borel sets are defined collectively as follows.

3.1.1 DEFINITION

Of all the sigma algebras of subsets of the real line R which include all the intervals, the smallest one (see Problem 3.1.12 for its existence) is called the *Borel class* and denoted $\mathscr{B}$. If a subset B of R happens to be a member of $\mathscr{B}$, B is called a *Borel set*.

The Borel class is known to contain practically any set in R that we may ever want to consider. Non-Borel sets are difficult to describe and need not concern us here. For some people "a Borel set" is synonymous with "practically any set." Such a viewpoint is entirely adequate in many cases.

3.1.2 DEFINITION

By a *random variable* we mean a real-valued function X defined on a probability space $(S, \mathscr{S}, P)$

$$X : S \to R$$

such that for any Borel set B the inverse image $X^{-1}(B)$ is a member of $\mathscr{S}$.

In the following example a probability question regarding X is answered by referring to the underlying probability space. The probability that X assumes a value in B is denoted by $P[X^{-1}(B)]$ or more simply by $P\{X \in B\}$.

Example 1   Three ordinary dice are thrown. Let X be the number of aces that appear. If B is a Borel set containing 0, 1, 2, but not 3, for example the interval $[-1, 2]$, determine the probability $P\{X \in B\}$.

Solution   The sample set of the underlying probability space consists of all the triples formed by I, II, III, IV, V, VI. The random variable X assumes only the values 0, 1, 2, and 3. For example, $X[(\text{II, II, V})] = 0$ and $X[(\text{V, I, I})] = 2$. Now the only way X can assume values in B is for X to assume the values 0, 1, or 2. If A_r denotes the event that exactly r aces appear, then

$$X^{-1}(B) = A_0 \cup A_1 \cup A_2 = A_3^c$$

Therefore

$$P\{X \in B\} = P(A_3^c) = 1 - P(A_3)$$
$$= 1 - (\tfrac{1}{6})^3$$
$$\doteq 0.995$$

Given a random variable X the probability value $P\{X \in B\}$ varies as B is varied. In other words we have a set function, which we shall denote by P_X, assigning to each member of the Borel class $\mathscr{B}$ a real number between 0 and 1. Further, it is not difficult to show (see Problem 3.1.13) that P_X satisfies the two axioms of probability measure, and thus we obtain $(R, \mathscr{B}, P_X)$ as a probability space by itself. Accordingly we give the following definition.

3.1.3 DEFINITION

Given a random variable X on a probability space $(S, \mathscr{S}, P)$ the set function P_X described by

$$P_X(B) = P[X^{-1}(B)]$$

for each Borel set B in the real line R is called the *probability distribution* of X. The probability space $(R, \mathscr{B}, P_X)$ is said to be *induced* by X from $(S, \mathscr{S}, P)$.

Both probability spaces $(S, \mathscr{S}, P)$ and $(R, \mathscr{B}, P_X)$ are mathematical models of the same random experiment, with the latter at a more simplified (or abstract) level than the former.

In a way the probability distribution P_X describes just how the "probability mass" is spread along the real line R according to X. If for example $P_X([a, b])$ is nearly 1, this means that X assumes values in the interval $[a, b]$ most of the time or that most of the probability mass is spread in this interval $[a, b]$ (or so we imagine anyhow). Furthermore, since $P_X(B) = P[X^{-1}(B)]$; that is, the amount of probability mass on B (a subset of R) is exactly the amount of probability mass on $X^{-1}(B)$ (a subset of S), we may visualize X as an agent that transfers the probability mass from S to R in a certain way.

Although P_X is in general difficult to describe since it is a set function, in some cases the entire probability mass is concentrated on just a few points so that the specification of the amount of probability mass at each of these points does completely describe P_X. We illustrate this by the following example.

Example 2   A coin with probability p of falling heads is tossed n times. Let X be the random variable that "counts" the number of heads in each sequence of n tosses. Determine the probability distribution P_X.

Solution   Since X has the range $0, 1, 2, \ldots, n$, the entire probability mass will be concentrated on just these $n + 1$ points. Therefore, if we determine

$$p_r = P\{X = r\} \qquad \text{for} \quad r = 0, 1, 2, \ldots, n$$

then for any Borel set B we have

$$P_X(B) = \sum_{r \in B} p_r$$

and P_X is completely determined. Now denoting by A_r the event of exactly r heads in the probability space of coin-tossing experiment, we calculate

$$p_r = P[X^{-1}(r)] = P(A_r) = C_r^n p_r (1 - p)^{n-r}$$

This result takes us to the following definition.

3.1.4 DEFINITION

A random variable X is said to have an (n, p) binomial distribution if P_X is such that

$$P_X(r) = C_r^n p_r (1 - p)^{n-r}$$

for $r = 0, 1, 2, \ldots, n$.

We point out that in $P_X(r)$ the r represents the Borel set consisting of the single number r, that is, to be strict we should have written $P_X(\{r\})$. Note also that

$$\sum_{r=0}^{n} P_X(r) = \sum_{r=0}^{n} C_r^n p^r (1 - p)^{n-r}$$
$$= [p + (1 - p)]^n = 1$$

so that all the probability mass is accounted for. A good way of visualizing a binomial distribution and other so-called "discrete" distributions is to place "mass points" along the real line. For example, the $(3, 0.3)$ binomial distribution has four mass points indicated in what we might call a *distribution diagram* (Fig. 3.1.2).

Figure 3.1.2   0   1   2   3   4

In Figure 3.1.2 each circle representing a certain amount of probability mass should be regarded as concentrated at the center of the circle regardless of size of the circle. The sizes of these circles are in proportion to the following table data of $(3, 0.3)$ binomial distribution:

$$P_X(0) = b(3, 0.3; 0) = 0.3430$$
$$P_X(1) = b(3, 0.3; 1) = 0.4410$$
$$P_X(2) = b(3, 0.3; 2) = 0.1890$$
$$P_X(3) = b(3, 0.3; 3) = 0.0270$$

Binomial distributions are one of the three most common types of (probability) distributions for random variables, the other two types being Poisson distributions and normal distributions.

3.1.5 DEFINITION

A random variable X is said to have a μ Poisson distribution if P_X is such that

$$P_X(k) = e^{-\mu} \frac{\mu^k}{k!} \qquad \mu > 0$$

for $k = 0, 1, 2, 3, \ldots$. Again we note that

$$\sum_{k=0}^{\infty} P_X(k) = e^{-\mu} \sum_{k=0}^{\infty} \frac{\mu^k}{k!}$$
$$= e^{-\mu} \cdot e^{\mu} = 1$$

so that all the probability mass is accounted for. The distribution diagram for μ Poisson distribution with $\mu = 3.5$ is roughly as in Fig. 3.1.3.

Figure 3.1.3

Note that $P_X(k)$ increases as k approaches μ and decreases as k passes and leaves μ behind; for an extremely large values of k, $P_X(k)$ is extremely small, in fact $P_X(k)$ approaches 0 as k approaches ∞. Figure 3.1.3 is based on the following table data of Poisson distribution for $\mu = 3.5$:

$$P_X(0) = p(3.5; 0) = 0.0302$$
$$P_X(1) = p(3.5; 1) = 0.1057$$
$$P_X(2) = p(3.5; 2) = 0.1850$$
$$P_X(3) = p(3.5; 3) = 0.2158$$
$$P_X(4) = p(3.5; 4) = 0.1888$$
$$P_X(5) = p(3.5; 5) = 0.1322$$
$$P_X(13) = p(3.5; 13) = 0.0001$$

The binomial distributions and the Poisson distributions are two typical examples of what is known as a discrete distribution.

3.1.6 DEFINITION

A random variable X is said to have a *discrete distribution* if there exists a sequence of real numbers $x_1, x_2, x_3, \ldots$ such that $P_X(x_i) > 0$ and

$$\sum_i P_X(x_i) = 1$$

In other words the entire probability mass is concentrated on mass points $x_1, x_2, x_3, \ldots$. In contrast to discrete distributions we have the so-called continuous distributions whereby the probability mass is spread smoothly along the real line or a portion of the real line.

3.1.7 DEFINITION

A random variable X is said to have a *continuous distribution* if P_X is such that for each singleton set $\{x\}$ we have

$$P_X(x) = 0$$

In other words the probability mass never occurs in "lumps" but rather is scattered out all over R. A good way of visualizing a continuous distribution is to draw a curve above R in such a way that the height of the curve "reflects" the amount of the mass spread directly below.

Figure 3.1.4 indicates a continuous distribution in which most probability mass is spread about the point 0 and no probability mass is found to the right of the point 2. The following example shows how casual a continuous distribution can be.

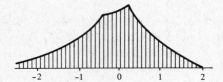

Figure 3.1.4

Example 3   If a dart is thrown with the aim to hit the center of a circular board of radius 1 ft and if X denotes the distance from the center to the point hit, then a likely distribution of P_X is indicated in Fig. 3.1.5 by the solid curve. A superior marksmanship would give rise to a distribution more heavily packed around the point 0 indicated by the dotted curve. In drawing various "density curves" to describe continuous distributions we shall always keep the area underneath the curve exactly 1.

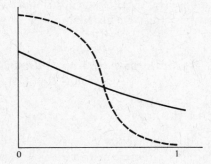

Figure 3.1.5

3.1.8 DEFINITION

Given a continuous random variable X, a nonnegative real-valued integrable function f is said to be a *density function* for P_X or simply for X if for any interval $[a, b]$

$$P([a, b]) = \int_a^b f(x)\, dx$$

In other words f is such that the probability mass on the interval $[a, b]$ is represented by the area underneath the graph of f over the interval $[a, b]$. Of course f has to be such that $\int_{-\infty}^{\infty} f(x)\, dx = 1$ so that the entire probability mass is accounted for. Any nonnegative real-valued function which is integrable over every interval and whose integral over the entire real line equals 1 can be used as a density function to describe some continuous distribution. We hasten to point out however that not every continuous distribution can be described with a density function (cf. Section 3.5 for further details).

3.1.9 DEFINITION

A random variable X is said to have a *uniform* (continuous) *distribution* over the interval $[a, b]$ if it has the density function specified as follows (see Fig. 3.1.6):

$$f(x) = \begin{cases} \dfrac{1}{b-a} & \text{for} \quad a \leq x \leq b \\ 0 & \text{for other } x \end{cases}$$

The probability value $P_X([c, d])$ is represented by the shaded area. Note that a density function need not be continuous as long as it is integrable. Also it is possible that two slightly different density functions describe the same continuous distribution; this is due to the fact that a function may be altered at a few isolated points without affecting its integral. A density function for P_X is often denoted by f_X.

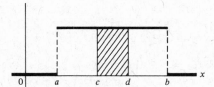

Figure 3.1.6

3.1.10 DEFINITION

A random variable X is said to have the *standard normal distribution* if

$$f_X(x) = \frac{1}{\sqrt{2\pi}} e^{-(1/2)x^2} \qquad \text{for} \quad -\infty < x < \infty$$

The graph of f_X (distribution diagram for P_X) is a bell-shaped curve symmetric with respect to $x = 0$ (see the solid curve in Fig. 3.1.7).

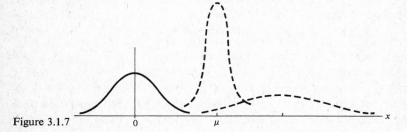

Figure 3.1.7

3.1.11 DEFINITION

A random variable X is said to have a (μ, σ) *normal distribution* if

$$f_X(x) = \frac{1}{\sigma\sqrt{2\pi}} \exp\left[-\frac{1}{2}\left(\frac{x-\mu}{\sigma}\right)^2\right] \quad \sigma > 0$$

for $-\infty < x < \infty$.

The graph of a (μ, σ) normal density function is similar to that of a standard normal density function except that the "hump" may be sharper if $\sigma < 1$ and the line of symmetry is at $x = \mu$ instead (see the tall dotted curve in Fig. 3.1.7). As σ gets larger the hump gets progressively flatter (see the other dotted curve in the figure).

The importance of normal distributions derives from the fact that many random variables we encounter have distributions that are *approximately* normal.

3.1 PROBLEMS

1. A rock is flung through a glass window. If X and Y denote respectively the amount and the number of pieces of the glass falling out, what kind of distributions do X and Y have, continuous or discrete? Sketch the distributions of X and Y.

2. It is quite possible for two distinct random variables to have identical distributions. If a coin with probability p of falling heads is tossed n times and X denotes the number of heads and Y the number of tails, show that X and Y have identical distributions if $p = \frac{1}{2}$. Hint: Show $P_X(r) = P_Y(r)$ for $r = 0, 1, 2, \ldots, n$.

3. In general, many different random variables may be associated with a given probability space. A needle is dropped on a table at random. Construct an appropriate sample set. What random variables can you consider?

4. Two fair coins are tossed; a \$5 bill is given for each head. Let X be the random variable "counting" the number of heads (0, 1, or 2), and let $Y = 5X$. Sketch the distributions for X and Y. Do the same for Z if $Z = X^3 - 5$.

5. In the experiment of tossing a biased coin 1000 times, if X and Y, respectively, count the number of heads and the number of tails, and if a new random variable is defined by $W = X + Y$, what kind of a distribution does W have?

6. If X counts the number of tosses until the first head appears in tossing a coin with probability p of falling heads, then $P_X(k) = pq^{k-1}$ for $k = 1, 2, 3, \ldots$, and X is said to have a *geometric distribution* with parameter p (or simply a p geometric distribution). Show that $\sum_k P_X(k) = 1$ and sketch the geometric distribution with $p = 0.9$. Use this sketch as an aid to determine the probability $P(X \geq 3)$.

7. From the fact that on the average 3.5 cars arrive at the toll-collecting station during the noon hour, the toll collector decides to use a Poisson distribution with parameter $\mu = 3.5$ to describe the random variable X denoting the number of cars arriving during the noon hour, that is:

$$P_X(k) = \frac{e^{-3.5}(3.5)^k}{k!} \quad \text{for} \quad k = 0, 1, 2, \ldots$$

Use the data for the Poisson distribution given just before Definition 3.1.6 to calculate the probability that at most 2 cars will arrive during the noon hour.

8. A random variable X is said to have a *triangular distribution* over the interval $[a, b]$ if the graph of its density function f_X is in the shape of an isosceles triangle with base $[a, b]$. Let X have a triangular distribution over $[-1, 1]$ and sketch the graph of f_X. Let the "cumulative" function F be defined by

$$F(x) = \int_{-\infty}^{x} f_X(t)\, dt$$

Sketch the graph of F. If you differentiate F, do you get back f_X? What basic theorem in calculus guarantees that you do?

9. Let X be discrete random variable having integer mass points $x_1, x_2, \ldots, x_n$ with corresponding probability masses $p_1, p_2, \ldots, p_n$. In other words, for n integers x_i we have

$$P_X(x_i) = p_i \quad \text{and} \quad \sum_{i=1}^{n} p_i = 1$$

Associated with X let $\bar{X}$ be a continuous random variable such that the probability mass p_i is spread uniformly over the interval $[x_i - \frac{1}{2}, x_i + \frac{1}{2}]$ for $i = 1, 2, \ldots, n$. In other words, $f_{\bar{X}}(x) = 0$ for all x except

$$f_{\bar{X}}(x) = p_i \quad \text{for} \quad x_i - \frac{1}{2} \leq x < x_i + \frac{1}{2}$$

with i ranging through $1, 2, \ldots, n$. Can you see that the distribution of $\bar{X}$ fairly approximates that of X? In fact

$$|P_X[a, b] - P_{\bar{X}}[a, b]| \leq \max_i p_i$$

for any interval $[a, b]$. The graph of $f_{\bar{X}}$ is known as the *histogram* of P_X. Can you still consider the histogram for a discrete X even when the mass points are not necessarily integers and in fact are closer together?

10. Let X denote the lifetime (in years) of a light bulb manufactured by a certain company in 1970. Although X is a random variable having a discrete distribution since the number of bulbs manufactured in that year is finite, the following continuous distribution

$$f_X(x) = 0.5e^{-0.5x} \qquad \text{for} \quad x \geq 0$$

is used as a reasonable approximation. Show that

$$\int_0^\infty f_X(x)\, dx = 1$$

and use the density function f_X to calculate the probability that a light bulb of the given description will have a lifetime of less than a year. In general, a random variable whose density function is of the form

$$f_X(x) = \lambda e^{-\lambda x} \quad \text{for} \quad x \geq 0$$

where λ is a positive constant is said to have an *exponential distribution*.

11. About the simplest of all random variables are those assuming only two values, 0 and 1. Such a random variable X seems to single out (hence indicate) a specific event $A = X^{-1}(1)$, and is thus called an *indicator function*. The indicator function of the event A is denoted I_A. Indicator functions may be used as building blocks for less simple random variables. For example if I_A and I_B, respectively, are indicator functions of event A and event B, and supposing $AB \neq \phi$, then the random variable

$$aI_A + bI_B$$

will assign to each outcome ω in the sample set S the real value

$$(aI_A + bI_B)(\omega) = aI_A(\omega) + bI_B(\omega)$$

Thus, for ω in AB the assigned value will be $a + b$, for ω in $A - B$ the assigned value will be a, and for ω in $B - A$ the assigned value will be b.

a. Show that $I_A + I_B = I_{A \cup B}$ for $AB = \phi$ and that

$$I_A \cdot I_B = I_{AB} \qquad \text{for any } A \text{ and } B$$

b. Show that if X is the random variable counting the number of heads in n tosses of a coin and if A_r is the event of exactly r heads, then

$$X = \sum_{r=0}^n rI_{A_r}$$

Show on the other hand that if E_i is the event of head at the ith toss, then

$$X = \sum_{i=1}^n I_{E_i}$$

so that if we let $X_i = I_{E_i}$, then

$$X = \sum_{i=1}^n X_i$$

Note that the indicator functions X_i serve as building blocks for X. In general, a random variable X is said to be *simple* if it is a finite linear combination of indicator functions, that is:

$$X = \sum_{i=1}^{n} a_i X_i$$

12 a. If $\mathscr{A}$ and $\mathscr{B}$ are both sigma algebras of subsets of S, show that $\mathscr{A}\mathscr{B}$ consisting of all subsets both in $\mathscr{A}$ and in $\mathscr{B}$ is again a sigma algebra of subsets of S. Likewise show that if $\{\mathscr{A}_i\}$ is a family of sigma algebra of subsets of S then $\bigcap_i \mathscr{A}_i$ is still a sigma algebra.

   b. Given a collection $\mathscr{C}$ of subsets of S, show the existence of both the largest and the smallest sigma algebras containing $\mathscr{C}$. Hint: Use (a) for (b). Let $\{\mathscr{A}_i\}$ be the family of all sigma algebras containing $\mathscr{C}$.

13. Let X be a random variable defined on a probability space $(S, \mathscr{S}, P)$. If for each Borel set $B \subset R$ we define

$$P_X(B) = P[X^{-1}(B)]$$

show that the set function P_X satisfies the following:
   i. $0 \leq P_X(B)$ and $P_X(R) = 1$.
   ii. $P_X(\bigcup_i B_i) = \sum_i P_X(B_i)$ for each disjoint sequence of Borel sets $B_1, B_2, B_3, \ldots$.
   Hint: Use $X^{-1}(\bigcup_i B_i) = \bigcup_i X^{-1}(B_i)$.

14. Let X, Y, Z be three random variables defined on $(S, \mathscr{S}, P)$ such that

$$Z \leq X + Y$$

Show that if $P(X > a) < \varepsilon$ and $P(Y > b) < \eta$, then

$$P(Z > a + b) \leq \varepsilon + \eta$$

3.2 DISTRIBUTION FUNCTIONS

The probability distribution P_X of a random variable X is supposed to describe precisely the tendency of X in assuming its various possible values. However, P_X is a set function, and set functions are in general hard to visualize. We shall, therefore, associate with each distribution P_X a real-valued function of a real variable F_X called the distribution function of X. The graph of F_X then can be regarded as a sort of geometric realization of P_X.

3.2.1 DEFINITION

Given a random variable X we define its *distribution function* F_X via its distribution P_X as follows:

$$F_X(x) = P_X((-\infty, x]) \qquad \text{for} \quad \text{each } x \in R$$

Example 1 If X assumes only two values $a < b$ with the corresponding probabilities p and q, determine the distribution function F_X.

Solution Clearly $F_X(x) = 0$ for $x < a$, $F_X(x) = p$ for $a \leq x < b$, and $F_X(x) = p + q = 1$ for $b \leq x$. Thus F_X is a step function that undergoes a jump of size p at $x = a$ and another jump of size q at $x = b$.

If we let $u(x)$ be the "unit step function" defined to be 0 for $x < 0$ and 1 for $0 \leq x$, then we may express F_X as

$$F_X(x) = pu(x - a) + qu(x - b)$$

We now examine the three basic properties of a distribution function which will tell us something about its general appearance (see Fig. 3.2.1).

3.2.1 PROPOSITION

F_X is a monotone increasing (or nondecreasing to be precise) function.

PROOF

If $x_1 < x_2$ then by the additive property of P_X we have

$$F_X(x_2) = P_X((-\infty, x_2]) = P_X((-\infty, x_1]) + P_X((x_1, x_2])$$

But $P_X((-\infty, x_1]) = F_X(x_1)$ and $P_X((x_1, x_2]) \geq 0$. Hence $F_X(x_1) \leq F_X(x_2)$.

3.2.2 PROPOSITION

$F_X(x)$ approaches 1 as x approaches ∞ and approaches 0 as x approaches $-\infty$. Therefore we may write $F_X(\infty) = 1$ and $F_X(-\infty) = 0$.

PROOF

The reading of this proof may be omitted by those who did not go through Proposition 1.5.1 carefully. Since F_X is monotone by the preceding proposition, we have

$$\lim_{x \to \infty} F_X(x) = \lim_{n \to \infty} F_X(n) = \lim_{n \to \infty} P_X((-\infty, n])$$

Now since $(-\infty, 1], (-\infty, 2], \ldots$ is a monotone increasing sequence of sets with the union equal to the real line R, we obtain in view of Proposition 1.5.1

$$\lim_{n \to \infty} P_X((-\infty, n]) = P_X((-\infty, \infty)) = P_X(R) = 1$$

Similarly, using $P_X(\phi) = 0$, we can show $\lim_{x \to -\infty} F_X(x) = 0$.

3.2.3 PROPOSITION

$F_X(x)$ is right-continuous at each x, that is, $\lim_{0 < \varepsilon \to 0} F_X(x + \varepsilon) = F_X(x)$ for any $x \in R$.

PROOF

This time we need Proposition 1.5.2. Again by the monotonicity of F_X we have

$$\lim_{0 < \varepsilon \to 0} F_X(x + \varepsilon) = \lim_{n \to \infty} F_X\left(x + \frac{1}{n}\right) = \lim_{n \to \infty} P_X\left(\left(-\infty, x + \frac{1}{n}\right]\right)$$

But the sets $(-\infty, x + 1/n]$ for $n = 1, 2, 3, \ldots$ form a decreasing sequence of sets with the intersection equal to $(-\infty, x]$, hence in view of Proposition 1.5.2 we have

$$\lim_{n \to \infty} P_X\left(\left(-\infty, x + \frac{1}{n}\right]\right) = P_X(-\infty, x] = F_X(x)$$

and this completes the proof.

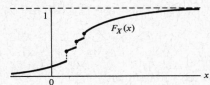

Figure 3.2.1

Now that we know that given a probability distribution P_X we can associate with it a monotone increasing right-continuous function F_X with $F_X(-\infty) = 0$ and $F_X(\infty) = 1$, we wonder whether given such a function F we can conversely find a probability distribution P_X such that its distribution function F_X is exactly the F given. The answer is yes, but the construction of the set function P_X is quite complicated (see Minibibliography: Royden). What is important for us here is the conclusion that P_X and F_X are two equally effective ways of describing the random nature of X.

Although the complete determination of P_X from F_X may be difficult, P_X may be determined easily to the extent that if B is some conveniently given interval rather than an unusual Borel set, then $P_X(B)$ can be calculated from F_X. If $B = (a, b]$, for example, then

$$P_X(a, b] = P_X(-\infty, b] - P_X(-\infty, a] = F_X(b) - F_X(a)$$

Example 2 The distribution function F_X of X is given as

$$F_X(x) = \frac{1}{\pi} \arctan x + \tfrac{1}{2}$$

From this determine $P_X(-\infty, 0]$ and $P_X 1(, \sqrt{3}]$.

Solution $P_X(-\infty, 0] = F_X(0) = 1/\pi \arctan 0 + \frac{1}{2} = \frac{1}{2}.$

$$P_X(1, \sqrt{3}] = F_X(\sqrt{3}) - F_X(1) = \frac{1}{\pi} (\arctan \sqrt{3} - \arctan 1)$$

$$= \frac{1}{\pi} \left(\frac{\pi}{3} - \frac{\pi}{4}\right) = \frac{1}{12}$$

We have seen that F_X is right-continuous everywhere; the necessary and sufficient condition under which F_X is left-continuous is stated in the following proposition.

3.2.4 PROPOSITION

At each x, F_X satisfies

$$\lim_{0 < \varepsilon \to 0} F_X(x - \varepsilon) = F_X(x) - P_X(x)$$

so that $F_X(x)$ is left-continuous if and only if $P_X(x) = 0$.

PROOF

Since F_X is monotone, $\lim_{0 < \varepsilon \to 0} F_X(x - \varepsilon) = \lim_{n \to \infty} F_X(x - 1/n)$, but $F_X(x - 1/n) = F_X(x) - P_X((x - 1/n, x])$. Consequently

$$\lim_{n \to \infty} F_X\left(x - \frac{1}{n}\right) = F_X(x) - \lim_{n \to \infty} P_X\left(\left(x - \frac{1}{n}, x\right]\right) = F_X(x) - P_X(x)$$

since the sets $(x - 1/n, x]$ for $n = 1, 2, 3, \ldots$ form a decreasing sequence of sets with intersection equal to $\{x\}$.

Propositions 3.2.3–4 together give rise to the following simple characterization of a continuous distribution in terms of its distribution function.

3.2.1 COROLLARY

P_X is a continuous distribution if and only if F_X is a continuous function.

We have defined a distribution function for a given random variable via its probability distribution; thus it is evident that we can associate a distribution function with a probability distribution without specific reference to the underlying random variable. A probability distribution can be conceived independently of random variables if it is understood to be nothing but a probability measure Q with domain $\mathscr{B}$, the Borel sets in the real line R. In other words, given a probability distribution Q (i.e., a probability measure on R) we can define the distribution function of Q by setting

$$F(x) = Q((-\infty, x])$$

Example 3 Let Q be the standard normal distribution so that for any Borel set B we have

$$Q(B) = \int_B \frac{1}{\sqrt{2\pi}} e^{-(1/2)x^2} \, dx$$

in particular,

$$Q([a, b]) = \frac{1}{\sqrt{2\pi}} \int_a^b e^{-(1/2)x^2} \, dx$$

Sketch the graph of the distribution function (usually denoted Φ) for Q.

Solution The graph of the normal distribution function

$$\Phi(z) = \frac{1}{\sqrt{2\pi}} \int_{-\infty}^z e^{-(1/2)x^2} \, dx$$

is a drawn S-shaped curve symmetric with respect to the point $(0, \frac{1}{2})$ as shown in Fig. 3.2.2.

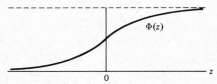

Figure 3.2.2

To check that

$$\lim_{z \to \infty} \Phi(z) = \frac{1}{\sqrt{2\pi}} \int_{-\infty}^\infty e^{-(1/2)x^2} \, dx = 1$$

consider somewhat artificially the product of two identical integrals

$$\left[\int_{-\infty}^\infty e^{-(1/2)x^2} \, dx \right]\left[\int_{-\infty}^\infty e^{-(1/2)y^2} \, dy \right] = \int_{-\infty}^\infty \int_{-\infty}^\infty e^{-(1/2)(x^2+y^2)} \, dy \, dx$$

which is exactly the volume V underneath the surface

$$f(x, y) = e^{-(1/2)(x^2+y^2)}$$

Now, using the usual technique of evaluating "volume of revolution" (draw a picture for "element of volume") we obtain

$$V = \int_0^\infty 2\pi r e^{-(1/2)r^2} \, dr = -2\pi e^{-(1/2)r^2} \Big|_0^\infty = 2\pi$$

Consequently,

$$\int_{-\infty}^\infty e^{-(1/2)x^2} \, dx = \sqrt{2\pi}$$

The values of $\Phi(z)$ can be found from standard tables of probability and statistics. For example

$$\Phi(1) = 0.5 + 0.3413 \cdots$$
$$\Phi(2) = 0.5 + 0.4772 \cdots$$
$$\Phi(3) = 0.5 + 0.4987 \cdots$$

The following example shows how the probability distribution of a random variable may sometimes be found only through calculation of its distribution function. This example may be omitted in the first reading.

Example 4 Exact arrival times of cars passing a toll-collecting station are marked daily between 12 noon and 6 P.M. A typical outcome resulting from a six-hour observation would be a finite sequence of numbers $\langle t_1, t_2, \ldots, t_n \rangle$ with $0 \leq t_1 < t_2 < \cdots < t_n \leq 6$. Let X_t be the random variable counting the number of cars passing before and up to the moment t, thus $X_t(\langle t_1, t_2, \ldots \rangle) =$ number of t_i's less than or equal to t. For example:

$$X_3(\langle 0.5, 1.2, 2.5, 3.0, 4.7 \rangle) = 4$$

Now, if it is known (or assumed) that X_t for each t in $[0, 6]$ has the λt Poisson distribution with $\lambda > 0$ a fixed constant, that is:

$$P\{X_t = k\} = P\{\langle t_1, t_2, \ldots \rangle : t_k \leq t < t_{k+1}\}$$
$$= e^{-\lambda t}(\lambda t)^k / k! \qquad \text{for} \quad k = 0, 1, 2, \ldots$$

and if Y is the random variable denoting the exact arrival time of the seventh car, what is the distribution of Y like?

Solution First we note that since the seventh car may not arrive within the time interval $[0, 6]$, Y is not quite defined on the entire sample set. For example, $Y(\langle 1, 2, 6 \rangle)$ is not defined since for the outcome $\langle 1, 2, 6 \rangle$ there is no arrival time of the seventh car. We shall therefore discard the time limit of six hours and make an ideal assumption that the cars will keep coming indefinitely. Our sample set then is the set of all monotone increasing sequences of nonnegative real numbers

$$S = \{\langle t_1, t_2, \ldots \rangle : 0 \leq t_1 < t_2 < \cdots \}$$

and the random variable X_t for each fixed $t > 0$ has the λt Poisson distribution. Y is now defined on the entire S since for each outcome $\langle t_1, t_2, \ldots \rangle$ we have

$$Y(\langle t_1, t_2, \ldots \rangle) = t_7$$

We will first determine the distribution function F for Y and thence obtain the density function f by differentiating F. Now for each fixed $t > 0$ we have

$$F(t) = P\{Y \leq t\} = P\{\langle t_1, t_2, \ldots \rangle : t_7 \leq t\}$$

But the event $E = \{\langle t_1, t_2, \ldots \rangle : t_7 \leq t\}$ can be split up into

$$E = \bigcup_{k=7}^{\infty} \{\langle t_1, t_2, \ldots \rangle : t_k \leq t < t_{k+1}\}$$

and each event $\{\langle t_1, t_2, \ldots \rangle : t_k \leq t < t_{k+1}\}$ is just the event that exactly k cars have passed up to the time t. Hence we have

$$F(t) = P(E) = \sum_{k=7}^{\infty} P\{X_t = k\}$$

$$= \sum_{k=7}^{\infty} \frac{e^{-\lambda t}(\lambda t)^k}{k!}$$

$$= 1 - \sum_{k=0}^{6} \frac{e^{-\lambda t}(\lambda t)^k}{k!}$$

Differentiating $F(t)$ with respect to t, we obtain after some simplification

$$f(t) = \frac{\lambda^7}{6!} t^6 e^{-\lambda t} \qquad t > 0$$

The fact that $f(t)$ above is a density function for Y is justified by the fundamental theorem of integral calculus, namely if $F(t)$ has a derivative $f(t)$ which is integrable then

$$\int_a^b f(t) \, dt = F(b) - F(a)$$

$$= P\{a \leq Y \leq b\}$$

It is a matter of a simple exercise in differential calculus to sketch the graph of the density function $f(t)$ (see Fig. 3.2.3); $t = 6/\lambda$ seems to be the peak time for arrivals of the seventh car. The reader may find it worthwhile to show more generally that if Y_r is the arrival time of the rth car then its density function is

$$f_r(t) = \frac{\lambda^r}{(r-1)!} t^{r-1} e^{-\lambda t} \qquad t > 0$$

This last density function belongs to the family of "gamma density functions."

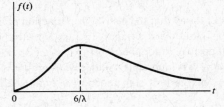

Figure 3.2.3

3.2.2 DEFINITION

A random variable X is said to have a (λ, r) *gamma distribution* if it has a density function $f(x)$ such that for $x > 0$

$$f(x) = \frac{\lambda^r}{\Gamma(r)} x^{r-1} e^{-\lambda x}$$

where $\lambda > 0$ and $r > 0$ are real constants and $\Gamma(r)$ is the so-called gamma function having the property $\Gamma(r) = (r-1)!$ for any positive integer r. (Details on the gamma function can be found in textbooks of advanced calculus; see also Problem 3.2.10.)

Example 4 or this section demonstrated that a density function may be obtained from the distribution function by differentiation, while Example 3 demonstrated that the distribution function may be obtained from a density function by integration.

3.2 PROBLEMS

1. Sketch the graphs for the distribution functions of the following probability distributions:

    a. An (n, p) binomial distribution.
    b. A p geometric distribution.
    c. A uniform distribution over $[a, b]$.
    d. A λ exponential distribution.

2. Let A be an event with probability p and X the indicator function of A (cf. Problem 3.1.11). Let $\bar{X}$ be the continuous approximation of X described in Problem 3.1.9. Sketch the distribution functions of X and $\bar{X}$ for comparison.

3. The distribution function of the one-point distribution at $x = 0$ is known as the *unit step function* denoted $u(x)$. Show that if X is a *simple random variable*, that is, discretely distributed over a finite set of points $x_1, x_2, \ldots, x_n$ with $P_X(x_i) = p_i$ and $\sum_i p_i = 1$, then the distribution function of X is given by

$$F_X(x) = \sum_{i=1}^{n} p_i u(x - x_i)$$

Graph $F_X(x)$ to see why such a function is called a *step function*.

4. It is not quite correct to say that the distribution function of a discrete distribution is always a step function. Consider for example a random variable X discretely distributed over the set of all positive rational numbers $x_1, x_2, x_3, \ldots$ (arranged in a certain order) such that $P_X(x_i) = (\frac{1}{2})^i$. Is it possible to describe F_X graphically? Can you describe F_X analytically in terms of the unit step function $u(x)$ introduced in the preceding problem?

5. If a random variable X has the distribution function

$$F_X(x) = \begin{cases} x + \frac{1}{2} & \text{for } -\frac{1}{2} \leq x \leq 0 \\ \frac{1}{4}x + \frac{1}{2} & \text{for } 0 \leq x \leq 2 \end{cases}$$

sketch the density function f_X from the graph of F_X. Then calculate f_X from F_X by formal differentiation. Note that the values of $f_X(x)$ at $x = -\frac{1}{2}$ and 2 may be arbitrarily (or conveniently) assigned.

6. If a random variable X has the density function

$$f_X(x) = \frac{2}{\pi}\sqrt{1 - x^2} \qquad \text{for } -1 \leq x \leq 1$$

sketch the distribution function F_X from the graph of f_X. Then calculate F_X from f_X by formal integration.

7. A dart is thrown at random at a circular board of radius 1 ft. Let X denote the distance measured in feet from the center of the board to the point hit. Find the distribution function F_X then the density function f_X by differentiating F_X. Could you have found f_X directly without going through F_X?

8. Each outcome of a certain experiment gives rise to a real number X. Before each performance of the experiment every onlooker is invited to choose an interval of unit length in which he bets the real number will occur. Where should you place your unit interval if you have the inside information that X has the distribution function

$$F(x) = \frac{1 - \cos x}{2}$$

for $0 \leq x \leq \pi$?

9. A particle moves back and forth without rest between the end points of the unit interval $[0, 1]$. As it moves from 0 to 1, it obeys the law $u = t^2$ where u is the distance between the point 0 and the particle and t is the time measured from the instant the particle leaves 0. As it moves back from 1 to 0, its motion obeys the law $v = t^3$ where v is the distance between the point 1 and the particle and t is the time measured from the instant the particle leaves 1. Determine whether the particle is more likely to be in the left half of the interval $[0, \frac{1}{2}]$ or the right half of the interval $[\frac{1}{2}, 1]$.

10. *Gamma function.* Let the gamma function be defined by

$$\Gamma(r) = \int_0^\infty t^{r-1}e^{-t}\, dt \qquad r > 0$$

a. Show with integration by parts, $u = e^{-t}$ and $dv = t^{r-1}\, dt$, that $\Gamma(r) = \Gamma(r + 1)/r$ and hence that $\Gamma(r + 1) = r\Gamma(r)$.

b. Show $\Gamma(1) = 1$ and, hence, that $\Gamma(n + 1) = n!$

c. Show $\Gamma(\frac{1}{2}) = \int_0^\infty {}^{-(1/2)}e^{-t}\, dt = \sqrt{\pi}$ by change of variables $t = u^2$. Recall also $\int_{-\infty}^\infty e^{-(1/2)x^2}\, dx = \sqrt{2\pi}$ from Section 3.2, Example 3.

3.3 TRANSFORMATIONS OF A RANDOM VARIABLE

If X is a random variable defined on a probability space $(S, \mathscr{S}, P)$ and g is a real-valued function of a real variable, then the composition $g(X)$ of g and X defined by

$$[g(X)](\omega) = g[X(\omega)] \qquad \text{for any} \quad \omega \in S$$

can be considered as a random variable on $(S, \mathscr{S}, P)$, provided that g satisfies a certain routine requirement.

3.3.1 DEFINITION

A mapping g of R into R is said to be a *Borel function* if for each Borel set B in the range R the inverse image $g^{-1}(B)$ is a Borel set in the domain R.

Looking at the diagram below, we see that if g is a Borel function then for any Borel set B in the range R of g we have

$$[g(X)]^{-1}(B) = X^{-1}[g^{-1}(B)] \in \mathscr{S}$$

so that $g(X)$ is indeed a random variable.

Now $g(X)$ can be thought of as a random variable obtained by transforming X by g. Most functions we encounter are Borel functions; in particular, all continuous functions are known to be Borel functions. This gives us a wide variety of ways in which to transform a given random variable. The following example illustrate how a $g(X)$ may arise in a natural context.

Example 1 After paying \$10 you are given a fair coin to toss until it falls heads. If you are paid \$3 for each toss you make, what is the probability of your ending up with a loss?

Solution If X denotes the number of tosses you make and Y denotes your net gain in dollars, then $Y = g(X) = 3X - 10$. In other words Y is a random variable obtained by transforming X by the Borel function $g(x) = 3x - 10$. The required probability is calculated as follows:

$$
\begin{aligned}
P(Y < 0) &= P(3X - 10 < 0) \\
&= P(X < \tfrac{10}{3}) \\
&= P(X = 1, 2, 3) \\
&= \tfrac{1}{2} + (\tfrac{1}{2})^2 + (\tfrac{1}{2})^3 \\
&= \tfrac{7}{8}
\end{aligned}
$$

The preceding example shows how a probability question regarding the derived random variable $Y = g(X)$ may be reduced to a probability question regarding the original random variable X. In general, given $Y = g(X)$ the probability distribution P_Y may be determined in terms of P_X and g. The following example shows how the density function f_Y may be determined from f_X if g is a simple linear function $g(x) = ax + b$.

Example 2 If X is a continuous random variable having the density function f_X and if $Y = aX + b$ with $a \neq 0$, relate f_Y to f_X.

Solution First we shall find F_Y, then by differentiating F_Y we obtain f_Y. For any fixed real number y, we have

$$F_Y(y) = P(Y \leq y) = P(aX + b \leq y)$$
$$= \begin{cases} P[X \leq (y-b)/a] = F_X[(y-b)/a] & \text{for} \quad a > 0 \\ P[X \geq (y-b)/a] = 1 - F_X[(y-b)/a] & \text{for} \quad a < 0 \end{cases}$$

Note that $P[X = (y-b)/a] = 0$ since X has a continuous distribution. Differentiating $F_Y(y)$ with respect to y, using the chain rule of differentiation, we obtain

$$f_Y(y) = \begin{cases} (1/a)f_X[(y-b)/a] & \text{for} \quad a > 0 \\ -(1/a)f_X[(y-b)/a] & \text{for} \quad a < 0 \end{cases}$$

In either case we have

$$f_Y(y) = (1/|a|)f_X[(y-b)/a]$$

The preceding example shows the effect of a "linear" transformation on the density function f_X. The resulting density function f_Y has a graph obtainable from that of f_X by merely shifting and stretching (or squeezing) the latter (try Problems 3.3.2–3.3.3). As an important special case of this we now show that every (μ, σ) normal density can be linearly transformed into the standard $(0, 1)$ normal density.

Example 3 If X is (μ, σ) normally distributed

$$f_X(x) = \frac{1}{\sigma\sqrt{2\pi}} \exp\left[-\frac{1}{2}\left(\frac{x-\mu}{\sigma}\right)^2\right]$$

where $\sigma > 0$ find f_Y for $Y = (X - \mu)/\sigma$.

Solution Since $Y = (1/\sigma)X - (\mu/\sigma)$, by the preceeding example we have

$$f_Y(y) = \frac{1}{1/\sigma} f_X\left(\frac{y + \mu/\sigma}{1/\sigma}\right)$$

A simple substitution of

$$x = \frac{y + \mu/\sigma}{1/\sigma} \quad \text{in } f_X(x)$$

gives

$$f_Y(y) = \frac{1}{\sqrt{2\pi}} e^{-(1/2)y^2}$$

In other words, Y is (0, 1) normal.

From the preceding example we see how a probability question concerning a (μ, σ) normal variable X may be converted into one concerning the standard normal variable Y. We demonstrate this by the following example.

Example 4 The test scores of a large calculus class have approximately a (60, 10) normal distribution. If grade A means 80 or better, what is the probability that a student arbitrarily picked from the class has grade A?

Solution Let X denote the test score. Since X is (60, 10) normally distributed, $Y = (X - 60)/10$ is (0, 1) normally distributed. Hence

$$P(X \ge 80) = P\left(\frac{X - 60}{10} \ge \frac{80 - 60}{10}\right)$$

$$= P(Y \ge 2) = 1 - \Phi(2)$$

but according to the table of values of standard normal distribution function Φ (see Section 3.2), $\Phi(2) \doteq 0.9772$. Therefore $P(X \ge 80) \doteq 0.0228$.

Since it may be cumbersome to memorize "density transformation" formulas such as the $f_Y(y) = (1/|a|)f_X[(y - b)/a]$ of Example 2 of this section, we shall now introduce a simple "differential technique" that will bring us these same formulas. This technique is primarily for those who are used to manipulating differentials from elementary calculus. Let X have a density function $f_X(x)$ and let $g(x) = ax + b$ transform X into $Y = aX + b$, then the amount of probability mass in the interval $(x, x + dx)$ for X is carried by g into the interval $(y, y + dy)$ or $(y + dy, y)$ depending on whether dy is positive or negative (see Fig. 3.3.1). Now the amount of probability mass in the interval $(x, x + dx)$ can be "differentially represented" by $f_X(x)\,dx$ while that in the corresponding interval for Y can be differentially represented by $f_Y(y)|dy|$ so that

$$f_Y(y)\,|dy| = f_X(x)\,dx$$

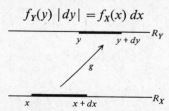

Figure 3.3.1

But then $|dy| = |a\,dx| = |a|\,dx$ since $y = g(x) = ax + b$; consequently

$$|a|f_Y(y)\,dx = f_X(x)\,dx$$

and hence

$$|a|f_Y(y) = f_X(x)$$

or

$$f_Y(y) = \frac{1}{|a|}f_X\left[\frac{(y-b)}{a}\right]$$

since $x = (y - b)/a$.

The result concerning linear transformations of a density function in Example 2 may be generalized by considering transformations g that are not necessarily linear, for example, $g(x) = x^2$ and arctan x. These useful generalizations are stated in the order of increasing generality as Propositions 3.3.1 and 3.3.2 below.

3.3.1 PROPOSITION

If X has a density function f_X and if $y = g(x)$ is a differentiable Borel function such that its derivative $g'(x)$ is continuous and positive (or negative, but not both), then the density function f_Y for $Y = g(X)$ may be obtained by

$$f_Y(y) = \frac{f_X(x)}{|g'(x)|}$$

for each y such that $x = g^{-1}(y)$ exists, otherwise $f_Y(y) = 0$.

Note, in particular, if $g(x) = ax + b$, the present formula reduces to the earlier

$$f_Y(y) = \frac{1}{|a|}f_X(x) \qquad \text{where} \quad x = g^{-1}(y) = \frac{y-b}{a}$$

PROOF

We give the proof for the case $g'(x) < 0$. Since g is a monotone decreasing function in this case, it takes one-to-one the whole real line on which X is distributed to some finite or infinite interval (a, b) in the real line on which Y is distributed. Clearly, for any y outside of (a, b), $f_Y(y) = 0$. As for y in (a, b), we have, proceeding as before,

$$F_Y(y) = P(Y \leq y) = P[g(X) \leq y]$$
$$= P[X \geq g^{-1}(y)]$$

since g is monotone decreasing. Hence,

$$F_Y(y) = 1 - P[X \leq g^{-1}(y)]$$

noting again that $P[X = g^{-1}(y)] = 0$ since X is continuous. Consequently,

$$F_Y(y) = 1 - F_X[g^{-1}(y)]$$

Differentiating $F_Y(y)$, again using the chain rule of differentiation, we obtain

$$f_Y(y) = -f_X[g^{-1}(y)] \cdot \frac{dx}{dy}$$

where we used dx/dy to denote the derivative of $x = g^{-1}(y)$ with respect to y. Noting that $dx/dy = 1/(dy/dx) = 1/g'(x)$ and also that $-dx/dy = |dx/dy|$ since $dx/dy < 0$, we obtain

$$f_Y(y) = \frac{f_X(x)}{|g'(x)|}$$

as required. In working out examples it helps to remember the above formula in the alternative form

$$f_Y(y) = f_X[g^{-1}(y)] \cdot \left| \frac{dx}{dy} \right|$$

The preceding formula may be derived "differentially" as follows. The amount of probability mass in the interval $(x, x + dx)$ is "differentially equal" to $f_X(x) \, dx$; an equal amount of probability mass is found in the corresponding interval $(y, y + dy)$ or $(y + dy, y)$, which is "differentially equal" to $f_Y(y) \, |dy|$ so that

$$f_Y(y) \, |dy| = f_X(x) \, dx$$

But

$$dy = g'(x) \, dx = \frac{dy}{dx} \cdot dx$$

consequently

$$f_Y(y) \left| \frac{dy}{dx} \right| dx = f_X(x) \, dx$$

By "canceling" dx, we obtain

$$f_Y(y) \left| \frac{dy}{dx} \right| = f_X(x)$$

or

$$f_Y(y) = f_X[g^{-1}(y)] \left| \frac{dx}{dy} \right|$$

Example 5 A double-headed arrow is mounted and spun freely a inches away from the origin of the real line calibrated in inches (see Fig. 3.3.2). As the arrow comes to a stop it will point out a point on the real line. Let Y denote the signed distance (positive to the right) of this point from the origin, and determine the distribution of Y.

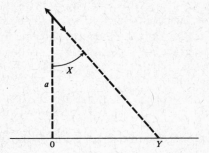

Figure 3.3.2

Solution Let the angle be measured as positive counterclockwise from the vertical line to the lower half of the arrow, and as negative clockwise. If X denotes this angle in radians, we may reasonably assume that X is uniformly distributed over the interval $[-\pi/2, \pi/2]$; in other words, $f_X(x) = 1/\pi$ in this interval. Clearly

$$Y = a \tan X$$

that is, Y is the result of transforming X by $y = g(x) = a \tan x$. Now if $y = a \tan x$, then $x = \arctan(y/a)$ and $dx/dy = a/(a^2 + y^2)$. Consequently, by Proposition 3.3.1 we have

$$f_Y(y) = f_X\left(\arctan \frac{y}{a}\right) \cdot \left| \frac{a}{a^2 + y^2} \right|$$

$$= \frac{1}{\pi} \cdot \frac{a}{a^2 + y^2} \qquad \text{for all } y$$

The preceding example gives us an occasion to introduce the following definition.

3.3.2 DEFINITION

A random variable X is said to have a *Cauchy distribution* if it has a density function of the form

$$f_X(x) = \frac{a}{\pi(a^2 + y^2)} \qquad a > 0$$

The graph of the Cauchy density function is somewhat similar to that of the normal density function; however, Cauchy density seems to spread out to $\pm\infty$ more liberally than normal density.

Some of the conditions we imposed on the function g in Proposition 3.3.1 are too restrictive. Suffice it to point out that even a simple function like $g(x) = x^2$ does not satisfy all the conditions of Proposition 3.3.1. Fortunately, most functions we encounter behave "locally" very much like the functions described in Proposition 3.3.1. Consider for example $Y = X^2$. Here the effect of $g(x) = x^2$ is to transfer the probability mass of X in the positive half line to the positive half line for Y and the probability mass of X in the negative half line to again the positive half line for Y. Thus, the positive half line for Y receives the probability mass "twice." In fact, assuming that X has a density function f_X we can find f_Y via F_Y in the usual way (see Problem 3.3.4):

$$f_Y(y) = \frac{f_X(\sqrt{y})}{2\sqrt{y}} + \frac{f_X(-\sqrt{y})}{2\sqrt{y}}$$

where $\sqrt{y}$ and $-\sqrt{y}$ appearing in $f_X(\cdot)$ are actually the two inverse images of y under the mapping $y = g(x) = x^2$, and the $2\sqrt{y}$'s appearing in the denominators are actually the two derivatives $|dx/dy|$ of the inverse functions $x = \sqrt{y}$ and $x = -\sqrt{y}$.

Functions such as x^2 and $\sin x$ may be characterized by "piecewise monotone differentiability."

3.3.3 DEFINITION

A mapping g of R into R is said to be *piecewise monotone differentiable* if there exists a chain of points

$$\cdots < c_{-2} < c_{-1} < c_0 < c_1 < c_2 < \cdots$$

in the domain R such that in each interval (c_n, c_{n+1}) the derivative $g'(x)$ exists, is continuous, and is never 0 (hence does not change sign). Any point x in the domain R other than the chain points c_n will be called a *regular point* of g. Also, any point y in the range R whose inverse image $g^{-1}(y)$ consists entirely of regular points in the domain R will be called a *regular point* of g.

For $y = g(x) = x^2$ the only nonregular points are $x = 0$ in the domain and $y = 0$ in the range. For $y = g(x) = \sin x$ the only nonregular points are $x = (2k + 1)(\pi/2)$, for $k = 0, \pm 1, \pm 2, \ldots$ and $y = \pm 1$. We will now state, but not prove, the following propositions.

3.3.2 PROPOSITION

If X has a density function f_X and if $y = g(x)$ is a piecewise monotone differentiable function, then the density function f_Y for $Y = g(X)$ may be obtained as follows: for each regular point y, we have

$$f_Y(y) = \sum_i \frac{f_X(x_i)}{|g'(x_i)|}$$

where the set $\{x_i\}$ is the inverse image $g^{-1}(y)$. If y is not a regular point or $g^{-1}(y)$ is empty, we simply set $f_Y(y) = 0$.

If x_i can be conveniently expressed as $g_i^{-1}(y)$ by means of piecewise inverse g_i^{-1} of g, then the above formula may be written alternatively as

$$f_Y(y) = \sum_i f_X[g_i^{-1}(y)] \cdot \left| \frac{dx_i}{dy} \right|$$

We illustrate the use of this formula by the following example.

Example 5 If X has the normal density

$$f_X(x) = \frac{1}{\sqrt{2\pi}} e^{-(1/2)x^2}$$

calculate the density for $Y = X^2$.

Solution Y is the result of transforming X by $g(x) = x^2$. Now if $y = g(x) = x^2$, then $x = \pm\sqrt{y}$; in other words

$$x_1 = g_1^{-1}(y) = \sqrt{y} \qquad \text{so that} \qquad \frac{dx_1}{dy} = \frac{1}{2\sqrt{y}}$$

and

$$x_2 = g_2^{-1}(y) = -\sqrt{y} \qquad \text{so that} \qquad \frac{dx_2}{dy} = -\frac{1}{2\sqrt{y}}$$

Hence, by Proposition 3.3.2, we have for each $y > 0$

$$f_Y(y) = f_X(\sqrt{y}) \left| \frac{1}{2\sqrt{y}} \right| + f_X(-\sqrt{y}) \left| -\frac{1}{2\sqrt{y}} \right|$$

$$= \frac{1}{2\sqrt{y}} \left(\frac{1}{\sqrt{2\pi}} e^{-(1/2)y} + \frac{1}{\sqrt{2\pi}} e^{-(1/2)y} \right)$$

$$= \frac{1}{\sqrt{2\pi}} y^{-(1/2)} e^{-(1/2)y}$$

The preceding example gives us an occasion to introduce the following definition.

3.3.4 DEFINITION

A random variable whose density function is of the form

$$f_X(x) = \frac{1}{2^{(n/2)}\Gamma(n/2)} x^{(n/2)-1} e^{-(x/2)} \qquad \text{for} \quad x > 0$$

is said to have a *chi-square distribution* with n degrees of freedom.

If we set $n = 1$, knowing that $\Gamma(\frac{1}{2}) = \sqrt{\pi}$ (see Problem 3.2.10) we obtain the density function of Example 5 of this section. Note also that chi-square density with n degrees of freedom is nothing but the $(\frac{1}{2}, n/2)$ gamma density. If we let $\lambda = \frac{1}{2}$ and $r = n/2$ in the gamma density function

$$\frac{\lambda^r}{\Gamma(r)} x^{r-1} e^{-\lambda x}$$

we obtain the n chi-square density function

$$\frac{1}{2^{(n/2)} \Gamma(n/2)} x^{(n/2)-1} e^{-(x/2)}$$

The χ^2 (chi-square) distributions are important for statisticians.

3.3 PROBLEMS

1. If X is (n, p) binomially distributed, that is:

$$P(X = r) = C_r^n p^r (1 - p)^{n-r} \quad \text{for} \quad r = 0, 1, 2, \ldots, n$$

and $Y = (X - np)/\sqrt{np(1 - p)}$, sketch the distributions of X and Y for $n = 2$ and $p = \frac{1}{2}$; do the same for $n = 3$ and $p = \frac{1}{3}$, also for $n = 4$ and $p = \frac{1}{4}$.

2. If X has a triangular distribution over the interval $[-0.1, 0.1]$ and if $Y = 10X + 2$, sketch the graphs of density functions f_X and f_Y, and compare them.

3. If X has a uniform distribution over the interval $[2, 3]$ and if $Y = 2X + 5$, sketch the graphs of density functions f_X and f_Y, and compare them.

4. If X has a density function $f_X(x)$ and if $Y = X^2$, calculate the density function $f_Y(y)$ by determining first the distribution function $F_Y(y)$.

5. If X has a uniform distribution over the interval $[0, 1]$, sketch the density functions f_Y for $Y = X^n$ and $n = 1, 2, 3, \ldots$ without calculations.

6. If X has a density function $f_X(x)$ and if $Y = X^3$, show that

$$f_Y(y) = \tfrac{1}{3} y^{-2/3} f_X(\sqrt[3]{y}) \quad \text{for all} \quad y$$

Assuming X has a uniform distribution over $[0, 2]$, sketch the graph of $f_Y(y)$.

7. Sketch an arbitrary f_X for X, then sketch $f_{|X|}$ by first establishing

$$f_{|X|}(y) = f_X(y) + f_X(-y) \quad \text{for} \quad y > 0$$

8. Let X have a triangular distribution over the interval $[-1, 1]$ and let g be

$$g(x) = \begin{cases} x & \text{for} \quad 0 \leq x \\ 0 & \text{otherwise} \end{cases}$$

Describe the distribution of $g(X)$ by sketching the graph of its distribution function.

9. Let Z be a random variable so continuously distributed that for any finite interval (a, b), $P_Z(a, b) > 0$ (think of a normally distributed Z, for example). If F_Z is the distribution function of Z and if X is uniformly distributed over the interval $[0, 1]$, show that $Y = F_Z^{-1}(X)$ has the distribution of Z. Hint: Show $F_Y(y) = F_Z(y)$ for all y.

3.4 RANDOM VECTOR, FAMILY OF RANDOM VARIABLES

So far we have considered only one numerical aspect at a time for a given random phenomenon, but actually there is no reason why we should not consider two or more simultaneously, especially if we are interested in comparing various numerical aspects of the given random phenomenon, or if we want to investigate the given random phenomenon more closely by examining several numerical aspects jointly. Indeed, we will now consider assigning to each outcome ω of a random experiment a set of numbers, $X_1(\omega)$, $X_2(\omega)$, ..., $X_n(\omega)$, which we shall regard as comprising the coordinates of a point (or vector) in the n-dimensional Euclidean space $R \times R \times \cdots \times R = R^n$. Imitating the definition of a random variable, we state the following definition.

3.4.1 DEFINITION

By an n-dimensional *random vector* we mean a mapping $\mathbf{X}$ of a probability space $(S, \mathscr{S}, P)$ into R^n such that whenever B is a Borel subset of R^n the inverse image $\mathbf{X}^{-1}(B)$ is a member of $\mathscr{S}$.

This definition has to be supplemented by the following well-anticipated definition.

3.4.2 DEFINITION

By the *Borel class* $\mathscr{B}^n$ of subsets of R^n we mean the smallest sigma algebra of subsets of R^n which includes all *rectangles* (a rectangle in R^n is a Cartesian product of intervals, $I_1 \times I_2 \times \cdots I_n$). A subset B of R^n is called a *Borel set* if B belongs to $\mathscr{B}^n$.

The idea behind $\mathscr{B}^n$ is simply that we want to consider the most economical family of subsets that includes all important subsets, notably the rectangles. Again, as far as we are concerned, a Borel set means practically any set.

Since a mapping $\mathbf{X}$ of S into R^n is clearly equivalent to a set of n mappings X_1, X_2, ..., X_n of S into R, we may write

$$\mathbf{X} = (X_1, X_2, \ldots, X_n)$$

and call X_i the ith component of $\mathbf{X}$. Since it can be shown that a mapping $\mathbf{X}: S \to R^n$ is a random vector (in the sense defined above) if and only if each component $X_i: S \to R$ is a random variable (in the sense defined earlier), technically a random vector is nothing but an indexed family of random variables all defined on a common probability space. However, it helps also to regard a random variable as a random vector of dimension 1.

Strictly speaking, a random vector $\mathbf{X}$ cannot be considered until the underlying probability space $(S, \mathscr{S}, P)$ is constructed, but in practice the consideration of $\mathbf{X}$ often follows immediately after the consideration of S; the study of $\mathbf{X}$ then helps to complete the definition of $\mathscr{S}$ and P, and the resulting $(S, \mathscr{S}, P)$ may then be viewed as the underlying probability space for $\mathbf{X}$. In other words, we need not hesitate to consider a random vector for a given random experiment before a definite probability space is constructed. Consider for instance the following example.

Example 1 (a) To each occurrence of an earthquake anywhere in the world assign a point (X, Y) on a map with X and Y respectively denoting the lattitude and the longitude of the epicenter. Here S consists of all points in the earth, the random vector $\mathbf{X}$ assigns to each point ω in S a certain point

$$\mathbf{X}(\omega) = (X, Y)(\omega) = [X(\omega), Y(\omega)]$$

in the map which is a portion of R^2.

(b) To each randomly selected college student we assign his height H, weight W, and intelligence quotient Q; the resulting random vector may be denoted

$$\mathbf{X} = (H, W, Q)$$

(c) To each sequence of 5 tosses of a die we assign the numbers of dots at the first, the third, and the fifth tosses. Denoting these by X_1, X_2, X_3, we have the random vector

$$\mathbf{X} = (X_1, X_2, X_3)$$

(d) To each daily observation of passing cars at a toll-collecting station we assign the total number of cars passed by 1 p.m., 2 p.m., and so on. We then have a random vector of any dimension n

$$\mathbf{X} = (X_1, X_2, \ldots, X_n)$$

Example 1(d) suggests the possibility of assigning to an outcome of a random experiment an infinite family of random variables. A random vector of an infinite dimension is known as a random process; random processes involving noncountably many random variables are studied in the theory of stochastic processes.

Since a random vector $\mathbf{X} = (X, Y)$ assumes vector values (or point values) in the plane $R_X \times R_Y$ (see Fig. 3.4.1), it is natural to ask: Given a Borel subset B in the plane, how likely is it that $\mathbf{X}$ takes a value in B? Now clearly the event "$\mathbf{X} \in B$" (read as $\mathbf{X}$ takes a value in B, *not* as $\mathbf{X}$ is a member of B) is represented in the underlying space S as $\{\omega: \mathbf{X}(\omega) \in B\}$ or $\mathbf{X}^{-1}(B)$, consequently we must set

$$P\{\mathbf{X} \in B\} = P[\mathbf{X}^{-1}(B)]$$

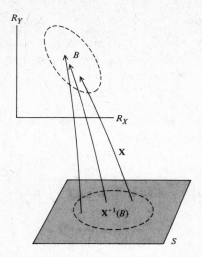

Figure 3.4.1

Since the probability value $P\{X \in B\}$ varies as B is varied, this gives rise to a set function P_X assigning to each Borel set B the probability

$$P_X(B) = P\{X \in B\}$$

More generally we consider the following definition.

3.4.3 DEFINITION

Given an n-dimensional random vector X on a probability space $(S, \mathscr{S}\ P)$, the set function P_X described by

$$P_X(B) = P[X^{-1}(B)]$$

for each Borel set B in R^n is called the *probability distribution* of X.

The resemblance of this definition to Definition 3.1.3 is striking. Also, like the distribution of a random variable, the distribution of a random vector can be shown to satisfy the axioms of probability measure. In other words, X will inevitably induce a probability space $(R^n, \mathscr{B}^n, P_X)$. Again it is helpful to visualize X as an agent transferring the probability mass from S to R^n. If X does not have too many components, we may write instead of P_X: $P_{(X, Y)}$, $P_{(X_1, X_2, X_3,)}$ or simply $P_{X, Y}$, P_{X_1, X_2, X_3}, and so on. P_{X_1, X_2, X_3} then may be called the *joint distribution* of the random variables X_1, X_2, and X_3. In this context we mention that any number of random variables X_1, X_2, ..., X_n, which are defined on one same probability space are said to be *jointly distributed*. Thus an n-dimensional random vector is synonymous to n jointly distributed random variables.

We push the analogy between P_X and P_X further and consider the following definition.

3.4.4 DEFINITION

A random vector $\mathbf{X}$ is said to have a discrete point $\mathbf{x} \in R^n$ if $P_{\mathbf{X}}(\mathbf{x}) > 0$ [though it would be more correct to write $P_{\mathbf{X}}(\{\mathbf{x}\}) > 0$]. $\mathbf{X}$ is said to have a *discrete distribution* if the discrete points account for all the probability mass, that is, there exist $\mathbf{x}_1, \mathbf{x}_2, \ldots,$ in R^n such that

$$\sum_i P_{\mathbf{X}}(\mathbf{x}_i) = 1$$

In contrast, $\mathbf{X}$ is said to have a *continuous distribution* if it does not have any discrete point, that is, for all $\mathbf{x}$ in R^n

$$P_{\mathbf{X}}(\mathbf{x}) = 0$$

Further, $\mathbf{X}$ is said to have an absolutely continuous distribution if $P_{\mathbf{X}}$ can be described in terms of a real-valued function $f(\mathbf{x})$ defined on R^n as follows: for any Borel set B in R^n

$$P_{\mathbf{X}}(B) = \int_B f(\mathbf{x}) \, d\mathbf{x}$$

where $\int_B \cdots d\mathbf{x}$ represents a multiple integral over the set B. For instance, for $n = 2$ we have simply the familiar double integral $\iint_B f(x, y) \, d(x, y)$. The real-valued function $f(\mathbf{x})$ is called a *density function* for $P_{\mathbf{X}}$ or $\mathbf{X}$ (see Example 3 of this section for a continuous distribution not having a density function).

Example 2 Whenever he feels like it, a self-employed taxi driver tosses a coin twice, then goes to work for as many hours as the number of heads. If X denotes the number of heads, and Y denotes the number of customers, what must the joint distribution of X and Y be like? Sketch the distribution for the random vector (X, Y).

Solution If $X = 0$, that is, the coin falls tails twice, the taxi driver declares a holiday so that $Y = 0$. Assuming the coin is fair, we have

$$P_{X, Y}(0, 0) = P(X = 0, Y = 0) = P(X = 0) = \tfrac{1}{4}$$

If $X = 1$, that is, the coin falls heads once, the number of customers tends to be smaller than if X is 2 since he works for only one hour. Supposing that the taxi driver serves 3 customers on the average during the first hour, and 5 customers during the two hours, we might assume that Y has a Poisson distribution with $\mu = 3$ when $X = 1$, and $\mu = 5$ when $X = 2$. For instance

$$P(Y = 10 \,|\, X = 1) = \frac{e^{-3}3^{10}}{10!}$$

$$P(Y = 0 \,|\, X = 2) = \frac{e^{-5}5^0}{0!}$$

In view of the sketch of Poisson distribution we had in Section 3.1 we can sketch the distribution of (X, Y) as in Fig. 3.4.2.

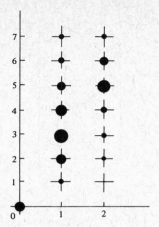

Figure 3.4.2

Note that

$$\sum_{k=0}^{\infty} P(X = 1, Y = k) = P(X = 1) = \tfrac{1}{2}$$

$$\sum_{k=0}^{\infty} (P(X = 2, Y = k) = P(X = 2) = \tfrac{1}{4}$$

Example 3 A rock is thrown through a glass window. Let X and Y, respectively, be the number of pieces and the exact amount in pounds of the glass falling out of the window. Does (X, Y) have a discrete or a continuous distribution? Sketch $P_{X, Y}$ if possible.

Solution Clearly $P_{X, Y}(x, y) = 0$ unless x is a positive integer since we have

$$P_{X, Y}(x, y) = P(X = x, Y = y) \leq P(X = x) = 0$$

for any x not a positive integer. But even if X is a positive integer, $P_{X, Y}(x, y) = 0$ since [recall the formula $P(AB) = P(A)P(B|A)$]

$$P(X = x, Y = y) = P(X = x)P(Y = y | X = x) = 0$$

for any exact value y. In other words, the probability mass is spread continuously along the vertical lines $x = 1, 2, 3, \ldots$. The probability mass of the amount $P(X = 3)$, say, is spread continuously along the vertical line $x = 3$ without a lump occurring at any specific point (see Fig. 3.4.3). Although we could consider a separate density function of one variable along each vertical line, it is impossible to conceive of a density function of two variables to describe the distribution of (X, Y). We have thus an example of a random vector which is continuous without being absolutely continuous.

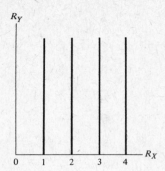

Figure 3.4.3

Example 4 A dart is thrown at a circular board of radius 1. Using the center of the board as the origin, let X and Y, respectively, be the abscissa and ordinate of the point hit. If we agree to describe the distribution $P_{X, Y}$ by the density function

$$f_{X, Y}(x, y) = \alpha(1 - \sqrt{x^2 + y^2}) \qquad \text{for} \quad x^2 + y^2 \leq 1$$

what must the constant α be? Sketch the graph of $f_{X, Y}$ and calculate the probability $P(\sqrt{X^2 + Y^2} \leq \frac{1}{2})$.

Solution The graph of $\alpha(1 - \sqrt{x^2 + y^2})$ is a cone of height α. Letting D be the unit disk around the origin, we must have

$$P_{X, Y}(D) = \iint_D \alpha(1 - \sqrt{x^2 + y^2}) \, dy \, dx = 1$$

Calculating the volume of the cone (see Fig. 3.4.4) directly without integration, we have

$$\tfrac{1}{3}\pi\alpha = 1 \quad \text{or} \quad \alpha = \frac{3}{\pi}$$

Consequently, we have

$$f_{X, Y}(x, y) = \frac{3}{\pi}(1 - \sqrt{x^2 + y^2})$$

We might say (X, Y) has a "triangular" distribution over the unit disk D. Next, letting B be the disk of radius $\frac{1}{2}$, we obtain

$$P(\sqrt{X^2 + Y^2} \leq \tfrac{1}{2}) = \iint_B f_{X, Y}(x, y) \, dx \, dy$$

$$= (\tfrac{1}{2})^2 \pi \cdot \frac{3}{2\pi} \cdot (1 + \tfrac{1}{3}) = \tfrac{1}{2}$$

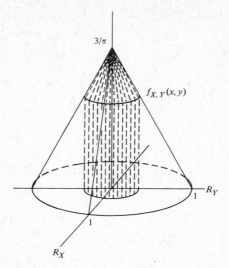

Figure 3.4.4

by simple geometric calculation. The marksmanship is better than "random" since for a uniform distribution over D the probability of $P_{X, Y}(B)$ would be only $\frac{1}{4}$.

Taken as a whole random vector $\mathbf{X}$ may be investigated in much the same way as a random variable $\mathbf{X}$, but once we start looking into the "stochastic relationship" that may exist among the component variables $X_1, X_2, \ldots, X_n$ of $\mathbf{X}$, we are treading on new ground. Although a systematic investigation of such relationship will not be given here (see Section 5.1, however), we will consider an important special case, namely that of *stochastic independence* of $X_1, X_2, \ldots, X_n$.

3.4.5 DEFINITION

Given a random variable X on a probability space $(S, \mathscr{S}, P)$, an event that X takes a value in a certain Borel set B is called an *X-event*. Such an event can be represented by

$$X^{-1}(B) = \{\omega \in S: X(\omega) \in B\}$$

Given a random vector $\mathbf{X} = (X_1, X_2, \ldots, X_n)$ we say that $\mathbf{X}$ has *independent components* or that $X_1, X_2, \ldots, X_n$ constitute an *independent class* of random variables if for any n given Borel sets $B_1, B_2, \ldots, B_n$ the X_i-events, $X_1^{-1}(B_1)$, $X_2^{-1}(B_2), \ldots, X_n^{-1}(B_n)$, always form an independent class of events.

As a simple illustration of Definition 3.4.5 consider the following example.

Example 5 Two dice are thrown. Let $X = 0$ or 1 according to whether the first die gives an even number or an odd number of dots. Likewise let $Y = 0$ or 1 according to the second die. Clearly, the random vector $\mathbf{X} = (X, Y)$ has independent components.

If X and Y are (stochastically) independent, the probability of an event described in terms of X and Y may be easily calculated from the probability of an X-event and the probability of a Y-event. For example, referring to Example 5 just now considered, and recalling the formula $P(AB) = P(A)P(B)$ for independent events A and B, we see that the probability of the event that $X = 0$ and $Y = 1$ can be calculated by

$$P(X = 0, \ Y = 1) = P(X = 0)P(Y = 1)$$

which is $\frac{1}{2} \times \frac{1}{2} = \frac{1}{4}$ if both dice are not loaded.

It is known in general that if a random vector has independent components then its probability distribution is completely determined by those of its components. We will consider two important special cases of this here, leaving the general case to Section 5.1.

3.4.1 PROPOSITION

If $\mathbf{X} = (X, Y)$ with X and Y independent and both having discrete distributions, say, $P_X(x_i) = p_i$ and $P_Y(y_j) = q_j$ for $i, j = 1, 2, 3, \ldots$, then $\mathbf{X}$ also has a discrete distribution, in fact

$$P_{\mathbf{X}}[(x_i, y_j)] = p_i q_j$$

PROOF

Since the range of X is $\{x_i\}_i$ and the range of Y is $\{y_j\}_j$, clearly the range of (X, Y) is $\{(x_i, y_j)\}_{i, j}$, which is a countable set. Now in view of independence of X and Y we have

$$\begin{aligned}
P_{\mathbf{X}}[(x_i, y_j)] &= P[X^{-1}(x_i) \cap Y^{-1}(y_j)] \\
&= P[X^{-1}(x_i)]P[Y^{-1}(y_j)] \\
&= P_X(x_i)P_Y(y_j) \\
&= p_i q_j
\end{aligned}$$

and

$$\sum_{i, j} P_{\mathbf{X}}[(x_i, y_j)] = \sum_{i, j} p_i q_j = \left(\sum_i p_i \right)\left(\sum_j q_j \right) = 1$$

3.4.2 PROPOSITION

If $\mathbf{X} = (X, Y)$ with X and Y independent and both having absolutely continuous distributions, say, with density functions $f_X(x)$ and $f_Y(y)$, then $\mathbf{X}$ also has an absolutely continuous distribution with density function given by

$$f_{X, Y}(x, y) = f_X(x)f_Y(y)$$

PROOF

Since a completely rigorous proof is impossible at this stage, we shall merely attempt to make the proposition plausible (see Proposition 5.2.2 for further consideration).

If X and Y are independent and both absolutely continuous with respective density functions f_X and f_Y, then for any pair of small intervals Δx in R_X and Δy in R_Y (see Fig. 3.4.5), we have

$$P(X \in \Delta x) = \int_{\Delta x} f_x(x)\, dx \doteq f_X(x^*)\, \Delta x$$

where x^* is a point in the interval Δx; likewise

$$P(Y \in \Delta y) \doteq f_Y(y^*)\, \Delta y$$

Consequently we have

$$P(X \in \Delta x,\ Y \in \Delta y) = P(X \in \Delta x)P(Y \in \Delta y)$$
$$\doteq f_X(x^*)f_Y(y^*)\, \Delta x\, \Delta y$$

In other words the probability mass on the rectangle $\Delta x \times \Delta y$ is approximately equal to the area of the rectangle $\Delta x\, \Delta y$ multiplied by a value of the function $f_X(x)f_Y(y)$ inside the rectangle $\Delta x \times \Delta y$. This leads us to suspect that the function of two variables $f_X(x)f_Y(y)$ may actually serve as the density function for $P_{X,\,Y}$. This turns out to be the case.

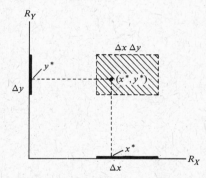

Figure 3.4.5

Example 6 A dart is thrown toward the origin of a coordinate plane. If (X, Y) denotes the coordinates of the point hit, and if we make the assumption that X and Y are independent and both $(0, \sigma)$ normally distributed, find the density function for (X, Y).

Solution From

$$f_X(x) = \frac{1}{\sigma\sqrt{2\pi}}\, e^{-(1/2\sigma^2)x^2} \quad \text{and} \quad f_Y(y) = \frac{1}{\sigma\sqrt{2\pi}}\, e^{-(1/2\sigma^2)y^2}$$

we obtain in view of Proposition 3.4.2

$$f_{X,\,Y}(x,\,y) = \frac{1}{\sigma^2(2\pi)}\exp\left[-\frac{1}{2\sigma^2}(x^2+y^2)\right]$$

The graph of this function is a surface of revolution of a normal curve.

Example 7 Suppose that $(X,\,Y)$ is a random vector with independent components and that X has a uniform distribution over the interval $[0, 0.2]$ and Y has an exponential distribution $f_Y(y) = 5e^{-5y}$ for $y \geq 0$. Calculate the probability that $Y \leq X$.

Solution We have $f_X(x) = 5$ for $0 \leq x \leq 0.2$ so that

$$f_{X,\,Y}(x,\,y) = f_X(x)f_Y(y) = 25e^{-5y}$$

for $0 \leq x \leq 0.2$ and $0 \leq y$. The entire probability mass for $P_{X,\,Y}$ is found on a narrow strip (Fig. 3.4.6). The event that $Y \leq X$ is represented by the portion of the plane below the diagonal line on which $Y = X$. The amount of probability mass on this portion of the plane is confined in a small triangle. Integrating the joint density function $f_{X,\,Y}(x,\,y)$ over this triangle, we obtain

$$P(Y \leq X) = \int_0^{0.2}\int_0^x 25e^{-5y}\,dy\,dx$$

$$= \frac{1}{e}$$

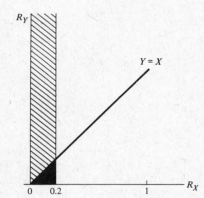

Figure 3.4.6

3.4 PROBLEMS

1. A dart is thrown at a square board. Let $\dot{X}$ and $\dot{Y}$ be respectively the measurements to the nearest hundredth of the distances X and Y from the left edge and the bottom edge of the board to the point hit by the dart. Does the random vector $(\dot{X},\,\dot{Y})$ have a discrete distribution? Does $(X,\,Y)$ have a continuous distribution? Do these two distributions approximate each other?

2. A dial is spun freely and allowed to come to a stop. Let X be the random vector denoting the exact position of the tip of the dial at rest. Does X have a continuous distribution, an absolutely continuous distribution?

3. An n-dimensional random vector $\mathbf{X}$ is said to have a *spherical distribution* if the amount of probability mass anywhere in R^n depends only on the distance from the origin and not on the direction relative to the origin. In other words, a spherical distribution is such that a rotation of the entire probability mass about the origin will not produce a new distribution. Determine which of the following two-dimensional random vectors have spherical distributions:

    a. (X, Y) have four discrete points, $(0, 1)$, $(1, 0)$, $(0, -1)$, $(-1, 0)$

    b. (X, Y) is uniformly distributed along the circle $x^2 + y^2 = 1$

    c. (X, Y) is uniformly distributed on the square $|x| + |y| \leq 1$

    d. (X, Y) has the density function

$$f(x, y) = \frac{1}{2\pi} e^{-(x^2 + y^2)/2}$$

4. An ordinary die, and a die loaded according to $p_i = ki$ where k is a constant and p_i is the probability that i dots occur, are tossed simultaneously. Calculate the probability that the loaded die gives more dots than the ordinary die.

5. Three random numbers X, Y, Z are drawn in succession from the interval $[0, 1]$. Determine the probability that $X \geq Y \geq Z$. Note that X, Y, Z are independent, each having a uniform distribution over $[0, 1]$.

6. Optional. A subset of the plane $R^2 = R \times R$ is called a *rectangular set* if it is of the form $I \times J$ where I and J are intervals in R. The smallest sigma algebra of subsets of R^2 which includes as members all rectangular sets is called the Borel class of subsets of R^2 and is denoted $\mathscr{B}^2$. Show that if A and B are Borel subsets of R then $A \times B$ is a Borel subset of R^2. In other words, $A, B \in \mathscr{B}^1$ implies $A \times B \in \mathscr{B}^2$.

7. Optional. Prove the following generalization of Problem 3.4.6. The Borel class $\mathscr{B}^n$ includes as its members all subsets of R^n of the form $B_1 \times B_2 \times \cdots \times B_n$ where the sets B_i are members of $\mathscr{B}^1$.

8. Optional. In view of Problem 3.4.7 show that $\mathscr{B}^n$ can be conceived of as the smallest sigma algebra of subsets of R^n that includes as members all "Borel rectangles" $B_1 \times B_2 \times \cdots \times B^n$ where the B_i are all Borel subsets of R^1. Thus we have an alternative definition for $\mathscr{B}^n$ besides Definition 3.4.2.

9. Propositions 3.4.1 and 3.4.2 are stated for random vectors of two dimensions. Can you state the similar propositions for random vectors of n dimensions?

10. Given a probability space $(S, \mathscr{S}, P)$ show that events $A_1, A_2, \ldots, A_n \in \mathscr{S}$ form an independent class if and only if their indicator functions $I_1, I_2, \ldots, I_n$ form an independent class.

3.5 SINGULAR DISTRIBUTION AND DECOMPOSITION OF DISTRIBUTIONS

Although most of the probability distributions we encounter in practice are either discrete, in which case the entire probability mass is concentrated on a number of points, or absolutely continuous, in which case the entire probability mass is spread out smoothly, in between these two types of distribution there is, at least in theory, another type in which the entire probability mass is neither quite concentrated as in the discrete case nor quite spread out as in the absolutely continuous case. This type of distribution arises when the entire probability mass is "stashed away" in some very "thin" set. Consider for example a two-dimensional random vector whose entire probability mass is spread uniformly on a line segment in the plane. The uniform spread along the line segment prevents the probability mass from concentrating at any point on the line segment or anywhere else. On the other hand the confinement of the probability mass along a line prevents any real-valued function of two real variables from serving as a density function since the double integral of such a function over the line segment (a set of zero area) is invariably 0. In short, the distribution in question is neither discrete nor absolutely continuous although we know at least that it is continuous since no probability mass is found at any single point. This particular example leads us to the following general definition. (Incidentally, throughout this section we shall denote a random vector by X instead of **X**. Thus, X will represent a random vector as well as a random variable).

3.5.1 DEFINITION

An n-dimensional random vector X (for $n = 1, 2, 3, \ldots$) is said to have a *singular continuous* distribution if P_X is such that for any x in R^n

$$P_X(x) = 0$$

and

$$P_X(E) = 1$$

for some "thin" set E in R^n. A Borel set E in R^n is said to be *thin* if the multiple integral over E of any real-valued function defined on R^n is 0, or as is said in measure theory, if E has zero (Lebesgue) measure.

Examples of singularly continuous n-dimensional random vectors for $n \geq 2$ are easy to find. For $n = 3$ consider for example a uniform distribution over the surface of a sphere and note that a surface has a zero volume in R^3 just as a line or a curve has a zero area in R^2. However, an example of a singularly continuous random variable ($n = 1$) is not so easy to find. The problem here is how to "stash away" the entire probability mass in a set of zero length without allowing the mass to "bulge up" at any single point. To see how difficult the problem is, let us try a finite set for E. If E consists of, say, three points x_1, x_2, x_3, then E certainly has a zero length, but unfortunately we cannot possibly distribute the probability mass of 1 to these three points without allowing the mass to bulge up at some of them, for if $P_X(x_1) = P_X(x_2) = P_X(x_3) = 0$ then $P(E) = 0 + 0 + 0 = 0$. If we try instead a countable set $E = \{x_1, x_2, x_3, \ldots\}$, for example the set of all rational numbers in [0, 1], we fail again since $P(x_i) = 0$ for all i will imply $P(E) = 0$ by the complete additivity of P. This leaves us only the non-countable sets of zero length as candidates for E. Now there are, to be sure, many such sets, but they were not known to us until G. Cantor (1845–1918) pointed out the first one. The set pointed out by Cantor is essentially what is left in the closed unit interval [0, 1] after a substantial portion of it has been removed. We shall describe it somewhat intuitively as follows. Take the closed interval [0, 1] to start with. Remove from [0, 1] the open middle third interval $(\frac{1}{3}, \frac{2}{3})$. From the remaining two intervals again remove their open middle thirds. As we keep on removing the successive middle thirds from the remaining intervals, we notice that there are points in [0, 1] that remain forever unaffected, for example the end points 0 and 1, also $\frac{1}{3}$ and $\frac{2}{3}$, and so on. These remaining points in [0, 1] constitute a set, which can be shown to be noncountable (one way is by "ternary expansions" of the reals) and of zero length (removed intervals add up to length 1). Named after its discoverer this set is known as Cantor's ternary set (or simply *the Cantor set*).

To construct a singularly continuous distribution in R we now take the Cantor set C and spread the entire probability mass "uniformly" on C; this can be done more precisely by constructing a distribution function that rises continuously from 0 to 1 over the interval [0, 1], but we shall not go into details (see Mini-bibliography: Royden). The important thing for us here is to be aware of the existence in much the same way of singularly continuous random variables as of the existence of singularly continuous random vectors of dimensions $n \geq 2$.

We have just pointed out that if the probability mass is allowed to "hide" in a set of zero measure (such as the Cantor set in R or a line segment in R^2), then it is impossible to find a density function to "describe" the hidden probability mass. Now, we can turn around and ask: What if no probability mass is hidden in a set of zero measure, can we then find a density function? The affirmative answer to this question is provided by the famous theorem of Radon-Nikodým (again see Royden).

3.5.1 THEOREM (Radon-Nikodým)
If X is an n-dimensional random vector ($n = 1, 2, 3, \ldots$) such that $P_X(E) = 0$ for every set of zero measure E in R^n, then there exists a real-valued function of n real variables f such that

$$P_X(B) = \int_B f(x)\, dx$$

for any Borel set B in R^n. In other words, X has a density function and so X must be absolutely continuous.

The existence of f is almost unique since if g is another such function, then $\int_B f(x)\,dx = P_X(B) = \int_B g(x)\,dx$ so that $\int_B [f(x) - g(x)]\,dx = 0$ for any and every Borel set B, and this means that f and g may differ only on a set of zero measure (see Royden).

The three types of distribution we have considered thus far involve the entire probability mass of a given random vector. However, it is possible, at least in theory, for a random vector to be distributed in such a way that part of its probability mass is concentrated on a number of points, another part of it stashed away in a thin set (or set of zero measure), and the rest spread out smoothly to admit a density function. Consider for example a three-dimensional random vector distributed in a sphere in R^3 in such a way that $\frac{1}{2}$ of the mass is distributed uniformly over the entire solid globe, $\frac{1}{4}$ of the mass is distributed uniformly over the spherical surface, $\frac{1}{8}$ of the mass is distributed uniformly over the equator circle, and finally $\frac{1}{16}$ each is distributed at the north and the south poles. It is natural to call such a distribution "mixed" since part ($\frac{1}{8}$) of the mass is distributed discretely, another part ($\frac{3}{8}$) is distributed singularly continuously, and the rest ($\frac{4}{8}$) absolutely continuously. In general, there are only these three ways of disposing the probability mass of a random vector. This assertion, originally due to H. L. Lebesgue (1875–1941), can be formulated within our context as follows.

3.5.2 THEOREM (Lebesgue Decomposition Theorem)

For any given n-dimensional random vector X there exist (1) a Borel set A in R^n such that $P_X(x) > 0$ for every x in A, (2) a Borel set E in R^n such that $P_X(x) = 0$ for every x in E and $\int_E g(x)\,dx = 0$ for every integrable g defined on R^n, and (3) an integrable function f defined on R^n such that

$$P_X(A) + P_X(E) + \int_{R^n} f(x)\,dx = 1$$

Note that $P_X(A)$ accounts for the amount of mass distributed discretely, $P_X(E)$ for the amount distributed singularly continuously, and the integral for the amount distributed absolutely continuously. In particular, we have a (purely) discrete distribution if $P_X(A) = 1$, in which case $E = \phi$ and $f = 0$; a continuous distribution if $A = \phi$, and a (purely) absolutely continuous distribution if $A = E = \phi$ so that the integral is equal to 1. Incidentally, although A is not stipulated to be finite or at most countable in the theorem, it cannot be otherwise on account of the restriction $P_X(A) \leq 1$. The detail of this can be worked out in Problem 3.5.1. The Lebesgue decomposition theorem makes the following definition meaningful.

3.5.2 DEFINITION

A random vector is said to have a *mixed* distribution if it is neither discretely distributed, singularly continuously distributed, nor absolutely continuously distributed.

The rest of this section is devoted to a brief study of distribution functions of random vectors. This portion may be omitted in the first reading.

Earlier we considered the distribution function F_X of a random variable X as a sort of geometric realization for the distribution P_X of the given random variable. We noted then that the continuity of P_X is reflected by the continuity of F_X. We might therefore expect that the same is true for n-dimensional random vector with $n \geq 2$. However, unfortunately perhaps, the singular part of the distribution may get in the way and we can end up with a random vector having a continuous distribution without having a continuous distribution function. The distribution function F_X of an n-dimensional random vector X is a real-valued function of n real variables, which we shall presently define after introducing the notion of hyperinterval (or box) in R^n. Hyperintervals are just as useful for studying functions on R^n as ordinary intervals are for studying functions on R^1.

Given $x = (x_1, x_2, \ldots, x_n)$ and $y = (y_1, y_2, \ldots, y_n)$ in R_n we write

$$x \prec y$$

if $x_i \leq y_i$ for $i = 1, 2, \ldots, n$. Now given $x \prec y$ we define the *closed hyperinterval* by

$$[x, y] = \{(t_1, t_2, \ldots, t_n): x_i \leq t_i \leq y_i\}$$

The *open hyperinterval* (x, y), *half-open hyperinterval* $(x, y]$ or $[x, y)$ can be similarly defined; in particular, we have

$$(-\infty, x] = \{(t_1, t_2, \ldots, t_n): -\infty < t_i \leq x_i\}$$

3.5.3 DEFINITION

By the *distribution function* of an n-dimensional random vector X we mean a real-valued function on R^n given by

$$F_X(x) = P_X(-\infty, x]$$

The three basic properties of F_X are
(1) Monotonicity (in the sense to be elaborated below);
(2) Two limits: $\lim_{x \to -\infty} F_X(x) = 0$ and $\lim_{x \to \infty} F_X(x) = 1$;
(3) Directed continuity: $\lim_{x_0 \prec x \to x_0} F_X(x) = F_X(x_0)$.

Monotonicity in the case of F_X where X is just a random variable means that

$$F_X(y) - F_X(x) \geq 0$$

whenever $x < y$, that is, a certain inequality holds for a combination of values of F_X evaluated at end points (or vertexes) of the interval $[x, y]$. Monotonicity of F_X where X is a random vector means something similar. Let us explain this step by step. First, by a *vertex* of a hyperinterval $[x, y]$ in R^n we mean a point $v = (t_1, t_2, \ldots, t_n)$ in $[x, y]$ such that each t_i is equal to either x_i or y_i. For example, $v = (y_1, x_2, \ldots, x_n)$ is a vertex, so are x and y in particular, and there are altogether 2^n vertexes of $[x, y]$. Next, we define the *index* of a vertex v by

$$i(v) = \text{the number of times in which } t_i = x_i$$

For example, $i(x) = n$ and $i(y) = 0$. Now it can be shown (but we omit the details) that

$$P_X(x, y] = \sum_v (-1)^{i(v)} F_X(v)$$

where the summation is made over v ranging through all 2^n vertexes of $[x, y]$. Note that for $n = 1$ we have exactly the familiar relation

$$P_X(x, y] = F_X(y) - F_X(x)$$

where X is a random variable. The reader can write out the similar relation for $n = 2$ as an exercise. And now, since $P_X(x, y] \geq 0$, this will impose the following inequality on the function F_X

$$\sum_v (-1)^{i(v)} F_X(v) \geq 0$$

for any hyperinterval $[x, y]$ of which v are vertexes. This inequality then is what we mean when we say that the distribution function F_X is *monotone increasing*. This monotonicity property of F_X goes much deeper than the superficial "coordinatewise monotonicity" (see Problems 3.5.6–3.5.7).

Example 1 Let the distribution P_X of a random vector $X = (X_1, X_2)$ be such that its entire probability mass is distributed normally on the vertical x_2-axis in R^2. Determine the distribution function F_X. Is P_X continuous? Is F_X continuous?

Solution P_X is clearly a singularly continuous distribution. Since no probability mass is to be found on the hyperinterval $(-\infty, x]$ for $x = (x_1, x_2)$ with $x_1 < 0$, we have

$$F_X(x_1, x_2) = 0 \quad \text{for} \quad x_1 < 0$$

On the other hand we have

$$F_X(x_1, x_2) = \Phi(x_2) \quad \text{for} \quad x_1 \geq 0$$

where Φ is the normal distribution function. Clearly then F_X is not continuous at $(0, x_2)$ for any x_2.

3.5 PROBLEMS

1. Let an n-dimensional random vector X be such that $P_X(x) > 0$ for every x in $A \subset R^n$. Show that the set A is at most countable. Hint: Enumerate all x in A such that $P_X(x) \geq \frac{1}{2}$, then all x such that $\frac{1}{2} > P_X(x) \geq \frac{1}{3}$, and so on.

2. The second hand of a clock takes $\frac{1}{4}$ sec to move from one stop to the next, then pauses for $\frac{3}{4}$ sec before advancing to the next stop. Let X be the random vector representing the position of the tip of the second hand. Describe the distribution of X.

3. In the preceding problem let X be the random variable representing the arc length traced out by the tip of the second hand measured from the 12 o'clock position. Sketch the graph of the distribution function F_X. Can you express F_X as the sum of two functions, one continuous and the other a step function?

4. Toss a fair coin. If it falls heads throw an ordinary die and count the number of dots; if it falls tails spin a wheel of fortune which gives a real number anywhere between 0 and 10. In either case let X denote the resulting number. Sketch the graph of the distribution function F_X.

5. Sketch, describe, or visualize the distribution functions for random vectors having the following distributions:
   a. A one-point distribution.
   b. A two-point distribution.
   c. A three-point distribution.
   d. A uniform distribution on the circle $x^2 + y^2 = 1$.
   e. A uniform distribution over the square $0 \leq x \leq 1$, $0 \leq y \leq 1$.

6. Show that the following function F does not have the monotonicity property in the sense described on page 124 although it is monotonous in x and y separately.

$$F(x, y) = \begin{cases} 0 & \text{if } x < 0 \text{ or } y < 0 \text{ or } x + y < 1 \\ 1 & \text{otherwise} \end{cases}$$

Hint: Take the unit square in the corner of the first quadrant and examine the values of F at the vertexes to obtain $\sum_v (-1)^{i(v)} F(v) < 0$.

7a. Given a distribution function $F_{X, Y}(x, y)$ show that

$$\lim_{x \to -\infty} F_{X, Y}(x, y) = \lim_{y \to -\infty} F_{X, Y}(x, y) = 0$$

   b. Show that if $F(x, y)$ has the monotonicity property postulated for distribution functions, then F is monotone in x and y separately.

3.6 BOREL FUNCTIONS OF A RANDOM VECTOR

If $\mathbf{X}$ is an n-dimensional random vector defined on a probability space $(S, \mathscr{S}, P)$ and $\mathbf{g}$ is an m-dimensional vector-valued function of an n-dimensional vector variable so that $\mathbf{X}$ and $\mathbf{g}$ may be joined as in the diagram then the composition $\mathbf{g}(\mathbf{X})$ of $\mathbf{g}$ and $\mathbf{X}$ defined by

$$[\mathbf{g}(\mathbf{X})](\omega) = \mathbf{g}[\mathbf{X}(\omega)] \qquad \text{for each} \quad \omega \text{ in } S$$

can be considered as an m-dimensional random vector provided that $\mathbf{g}$ satisfies a certain routine requirement.

3.6.1 DEFINITION

A mapping $\mathbf{g}$ of R^n into R^m is said to be a *Borel function* if for each Borel set B in R^m the inverse image $\mathbf{g}^{-1}(B)$ is a Borel set in R^n.

The notation $\mathbf{g}(\mathbf{X})$ representing the new random vector derived from $\mathbf{X}$ via $\mathbf{g}$ encompasses a wide variety of situations. These are Cases I, II, III, IV to be considered below. Case II has some important ramifications, among which is the notion of convolution discussed in details (these details are not needed in the sequel, however). Cases III and IV may be omitted from the first reading.

Case I

If $n = m = 1$, the random vector $\mathbf{g}(\mathbf{X})$ reduces to the earlier random variable $g(X)$ of Section 3.3.

Case II

If $n \geq 2$ and $m = 1$, $\mathbf{g}(\mathbf{X})$ is still a random variable, conveniently denoted as $g(\mathbf{X})$ or $g(X, Y)$ if $n = 2$; in particular if g is simply a "sum function," $g(x, y) = x + y$, then we may write $g(X, Y)$ simply as $X + Y$.

As might be expected, the basic problem concerning the random vector $\mathbf{Y} = \mathbf{g}(\mathbf{X})$ is that of determining the distribution of $\mathbf{Y}$ from the distribution of $\mathbf{X}$ according to the specific $\mathbf{g}$ involved. We shall illustrate this by a few concrete examples.

Example 1 Let the random vector (U, V) be equally distributed among the three points $(1, 0), (0, 1)$, and $(2, 2)$ (Fig. 3.6.1). Determine the distribution of Y if $Y = U + V$.

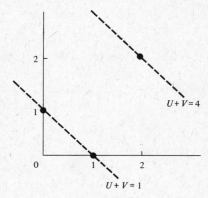

Figure 3.6.1

Solution From the figure it is clear that the only possible values for Y are 1 and 4, and these with probabilities $\frac{2}{3}$ and $\frac{1}{3}$.

Example 2 The velocity of a gas molecule is a random vector $\mathbf{V} = (X, Y, Z)$ with independent components each having a $(0, \sigma)$ normal distribution. Let S be the speed of such a gas molecule

$$S = \sqrt{X^2 + Y^2 + Z^2}$$

Determine the distribution of S by finding its density function.

Solution Here we have a random vector (X, Y, Z) having the density function (cf. Example 3.4.6).

$$f(x, y, z) = f(x)f(y)f(z) = \left(\frac{1}{\sigma\sqrt{2\pi}}\right)^3 \exp\left[-\frac{1}{2\sigma^2}(x^2 + y^2 + z^2)\right]$$

to be transformed by the Borel function

$$g(x, y, z) = \sqrt{x^2 + y^2 + z^2}$$

which sends points on the sphere of radius s in R^3 to the point s in R^1. Thus the probability mass in the shell with inner radius s and outer radius $s + ds$ is sent to the interval $(s, s + ds)$. Consequently, noting that the volume of the shell is $4\pi s^2\, ds$ and the length of the interval is ds, we have ("differentially" speaking)

$$\left(\frac{1}{\sigma\sqrt{2\pi}}\right)^3 \exp\left(-\frac{1}{2\sigma^2} s^2\right) 4\pi s^2\, ds = f_s(s)\, ds$$

After simplification we have

$$f_s(s) = \sqrt{\frac{2}{\pi}} \frac{s^2}{\sigma^3} \exp\left(-\frac{1}{2\sigma^2} s^2\right) \qquad \text{for} \quad s > 0$$

where f_s is known as the *Maxwell density function* in statistical mechanics.

As an important special case of $g(\mathbf{X})$ we consider the case where $\mathbf{X} = (X_1, X_2, \ldots, X_n)$ is a random vector having independent components and g is the sum function, $g(x_1, x_2, \ldots, x_n) = x_1 + x_2 + \cdots + x_n$ so that

$$Y = g(\mathbf{X}) = X_1 + X_2 + \cdots + X_n$$

and Y is just a *sum of independent random variables*. Now as long as g is kept unchanged the distribution of Y depends only on the distribution of $\mathbf{X}$, but the distribution of $\mathbf{X}$ is completely determined by the distributions of $X_1, X_2, \ldots, X_n$ in view of the stochastic independence of these random variables; consequently P_Y must depend on $P_{X_1}, P_{X_2}, \ldots, P_{X_n}$ alone. Accordingly, we consider the following definition.

3.6.2 DEFINITION

If $X_1, X_2, \ldots, X_n$ are jointly distributed independent random variables, and if

$$Y = X_1 + X_2 + \cdots + X_n$$

then the distribution of Y is called the *convolution* of the distributions of X_1, $X_2, \ldots, X_n$. We express this relationship by writing

$$P_Y = P_{X_1} * P_{X_2} * \cdots * P_{X_n}$$

Example 3 Find the convolution $P_X * P_Y$ if P_X and P_Y are given as follows:

$$P_X(1) = \tfrac{1}{3} \qquad P_X(2) = \tfrac{2}{3}$$
$$P_Y(2) = \tfrac{2}{5} \qquad P_Y(3) = \tfrac{3}{5}$$

Solution First we find $P_{X, Y}$ by

$$P_{X, Y}(1, 2) = P_X(1)P_Y(2) = \tfrac{1}{3} \times \tfrac{2}{5} = \tfrac{2}{15}.$$

Likewise

$$P_{X, Y}(1, 3) = \tfrac{3}{15}$$
$$P_{X, Y}(2, 2) = \tfrac{4}{15}$$
$$P_{X, Y}(2, 3) = \tfrac{6}{15}$$

Consequently (see Fig. 3.6.2)

$$(P_X * P_Y)(3) = P_{X, Y}(1, 2) = \tfrac{2}{15}$$
$$(P_X * P_Y)(4) = P_{X, Y}(1, 3) + P_{X, Y}(2, 2) = \tfrac{7}{15}$$
$$(P_X * P_Y)(5) = P_{X, Y}(2, 3) = \tfrac{6}{15}$$

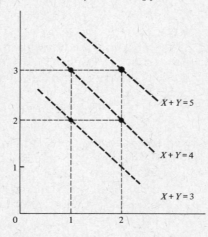

Figure 3.6.2

Example 4 Find $P_W = P_X * P_Y * P_Z$ if X has a two-point distribution at $x = 5$ and $x = 10$ with

$$P_X(5) = \tfrac{1}{3}$$
$$P_X(10) = \tfrac{2}{3}$$

and Y and Z are both uniformly distributed over the interval $[-1, 1]$.

Solution Writing $W = X + (Y + Z)$, we see

$$P_W = P_X * P_{Y+Z}$$

since quite obviously X and $Y + Z$ are independent. But

$$P_{Y+Z} = P_Y * P_Z$$

hence turning to $P_Y * P_Z$, we first determine $P_{Y,Z}$ (see Fig. 3.6.3) via

$$f_{Y,z}(y, z) = f_Y(y)f_Z(z) = \tfrac{1}{2} \cdot \tfrac{1}{2} = \tfrac{1}{4}$$

for $-1 \leq y \leq 1$, $-1 \leq z \leq 1$.

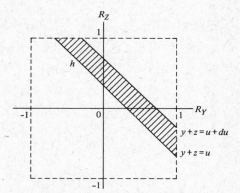

Figure 3.6.3

Now the shaded slanting strip between the lines $y + z = u$ and $y + z = u + du$ will be mapped to the interval $[u, u + du]$ in the real line R_U where $U = Y + Z$, transferring the mass $(\tfrac{1}{4}) \cdot h \cdot (du/\sqrt{2})$ where h denotes the length of the shaded strip and $du/\sqrt{2}$ is the width of the shaded strip. Consequently

$$\frac{1}{4} h \frac{du}{\sqrt{2}} = f_{Y+Z}(u) \, du$$

so that

$$f_{Y+Z}(u) = \frac{1}{4\sqrt{2}} h$$

Since h is longest when $u = 0$ and decreases linearly as u tends to ± 2, we see that $Y + Z$ has a triangular distribution over the interval $[-2, 2]$.

Finally since X takes only two values 5 and 10, adding $Y + Z$ to X will merely smear these two point masses into two triangular densities (see Fig. 3.6.4).

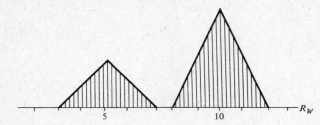

Figure 3.6.4

The following two propositions are useful in determining the convolutions of two discrete distributions and two absolutely continuous distributions respectively.

3.6.1 PROPOSITION

If P_X and P_Y are such that

$$\sum_{i=-\infty}^{\infty} P_X(i) = 1 \quad \text{and} \quad \sum_{j=-\infty}^{\infty} P_Y(j) = 1$$

then

$$(P_X * P_Y)(k) = \sum_{i=-\infty}^{\infty} P_X(i)P_Y(k-i)$$

PROOF

For each integer k we have

$$P(X + Y = k) = \sum_{i=-\infty}^{\infty} P(X = i, Y = k - i)$$

$$= \sum_{i=-\infty}^{\infty} P(X = i)P(Y = k - i)$$

$$= \sum_{i=-\infty}^{\infty} P_X(i)P_Y(k - i)$$

needless to say, one could also establish likewise

$$(P_X * P_Y)(k) = \sum_{j=-\infty}^{\infty} P_X(k - j)P_Y(j)$$

3.6.2 PROPOSITION

If P_X and P_Y are absolutely continuous with density functions $f_X(x)$ and $f_Y(y)$, then $P_X * P_Y$ is absolutely continuous with density function

$$f_{X+Y}(v) = \int_{-\infty}^{\infty} f_X(x)f_Y(v - x)\, dx$$

PROOF

We first determine the distribution function F_{X+Y}. For each real v, we have

$$F_{X+Y}(v) = P(X + Y \leq v) = P_{X,Y}(Q_v)$$

where $Q_v = \{(x, y): x + y \leq v\}$ as shown in Fig. 3.6.5. Now remembering that X and Y are independent, we have

$$P_{X,Y}(Q_v) = \iint_{Q_v} f_{X,Y}(x, y) \, d(x, y)$$

$$= \int_{-\infty}^{\infty} \left(\int_{-\infty}^{v-x} f_X(x)f_Y(y) \, dy \right) dx$$

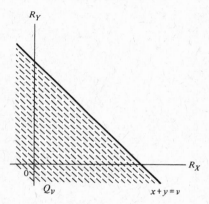

Figure 3.6.5

Differentiating with respect to v, we obtain

$$f_{X+Y}(v) = \int_{-\infty}^{\infty} \left(\frac{d}{dv} \int_{-\infty}^{v-x} f_X(x)f_Y(y) \, dy \right) dx$$

$$= \int_{-\infty}^{\infty} f_X(x)f_Y(v - x) \, dx$$

Likewise

$$f_{X+Y}(v) = \int_{-\infty}^{\infty} f_X(v - y)f_Y(y) \, dy$$

Example 5 Show that the convolution of two Cauchy distributions

$$f_X(x) = \frac{1}{\pi}(1 + x^2) \quad \text{and} \quad f_Y(y) = \frac{1}{\pi}(1 + y^2)$$

is again a Cauchy distribution (see Definition 3.3.2).

$$f_{X+Y}(v) = \frac{2}{\pi}(2^2 + v^2)$$

Solution By Proposition 3.6.2 we have

$$f_{X+Y}(v) = \int_{-\infty}^{\infty} \frac{1}{\pi(1+x^2)} \frac{1}{\pi[1+(v-x)^2]} dx$$

Integrating by partial fractions we obtain

$$\frac{1}{\pi^2(4+v^2)} \int_{-\infty}^{\infty} \left[\frac{(2/v)x+1}{1+x^2} - \frac{(2/v)x-3}{1+(v-x)^2}\right] dx = \frac{2}{\pi(2^2+v^2)}$$

where besides change of variables we also took advantage of the fact that the integral of an odd function vanishes.

Example 6 Show that the convolution of two λ exponential distributions is a $(\lambda, 2)$ gamma distribution (see Definition 3.2.2).

Solution Let X, Y be independent and

$$\left.\begin{array}{l} f_X(x) = \lambda e^{-\lambda x} \\ f_Y(y) = \lambda e^{-\lambda y} \end{array}\right\} \quad \text{for} \quad x > 0, \ y > 0$$

then by Proposition 3.6.2, we have, noting that $f_X(x) = 0$ for $x < 0$ and $f_Y(y) = 0$ for $y < 0$,

$$f_{X+Y}(v) = \int_0^v f_X(x)f_Y(v-x)\, dx$$

$$= \int_0^v \lambda e^{-\lambda x} \lambda e^{-\lambda(v-x)}\, dx$$

$$= \lambda^2 \int_0^v e^{-\lambda x}\, dx$$

$$= \lambda^2 e^{-\lambda v}\, v$$

which is the $(\lambda, 2)$ gamma density function. More generally, the r-fold convolution of λ exponential distributions is the (λ, r) gamma distribution (see Problem (3.6.10).

Proposition 3.6.2 impels us to the following definition.

3.6.3 DEFINITION

Given two (density) functions $f(x)$ and $g(y)$, their convolution $f * g$ is defined by

$$(f * g)(v) = \int_{-\infty}^{\infty} f(x)g(v-x)\, dx$$

$$= \int_{-\infty}^{\infty} f(v-y)g(y)\, dy$$

Although the computation is quite tedious, it can be shown that the convolution of two normal density functions is a normal density function. Specifically it can be shown that the convolution of $(\mu_1, \sigma_1), (\mu_2, \sigma_2), \ldots, (\mu_n, \sigma_n)$ normal densities is the $(\mu_1 + \mu_2 + \cdots + \mu_n, \sqrt{\sigma_1^2 + \sigma_2^2 + \cdots + \sigma_n^2})$ normal density. The investigation of probability distribution of the sum of independent random variables via convolution of density functions may lead to complicated integrals. However, in Chapter 6 we shall consider another means of describing probability distributions, namely the *characteristic functions* (Definition 6.3.2). Unlike density functions these functions may be convoluted by ordinary multiplication, and thus serve as a better tool for investigating the probability distributions of sums of independent random variables.

Case III

If $n = m \geq 2$, $\mathbf{Y} = \mathbf{g}(\mathbf{X})$ is a bona fide random vector, and we shall be concerned with the determination of the density function of $\mathbf{Y}$ in terms of the density function of $\mathbf{X}$ assuming that $\mathbf{g}$ is such that the absolute continuity of $\mathbf{X}$ is carried over to $\mathbf{Y}$. Note that if $\mathbf{g}$ is, for example, a constant function, then $\mathbf{Y}$ will have a one-point discrete distribution no matter what kind of distribution $\mathbf{X}$ has. A somewhat broad condition under which $\mathbf{Y}$ inherits absolute continuity from $\mathbf{X}$ is formulated in terms of the Jacobian derivative of $\mathbf{g}$, which we now introduce.

3.6.4 DEFINITION

A mapping $\mathbf{g}$ of R^n into R^n is said to be *differentiable* (or of class C^1) if all the partial derivatives of the following component functions of $y = g(x)$ are continuous:

$$y_1 = g_1(x_1, x_2, \ldots, x_n)$$
$$y_2 = g_2(x_1, x_2, \ldots, x_n)$$
$$\vdots$$
$$y_n = g_n(x_1, x_2, \ldots, x_n)$$

that is, $\partial y_j / \partial x_i$ exists and is continuous in $x = (x_1, x_2, \ldots, x_n)$ for each pair of $i = 1, 2, \ldots, n$, and $j = 1, 2, \ldots, n$. The *Jacobian derivative* of $\mathbf{g}$ at x is defined in terms of the following $n \times n$ determinant.

$$g(x) = \left| \left(\frac{\partial y_j}{\partial x_i} \right)_{i, j} \right|$$

Clearly if $\mathbf{g}$ is a constant mapping, then $g_1, g_2, \ldots, g_n$ are all constant functions and hence all partial derivatives are 0, so that $\mathbf{g}'(x) = 0$. But, needless to say, $\mathbf{g}'(x)$ can easily vanish without $\mathbf{g}$ being a constant mapping.

The following proposition which guarantees the absolute continuity of $\mathbf{Y}$ as well as provides a formula for calculation of the density function $f_{\mathbf{Y}}$ is analogous to Proposition 3.3.1 and stated here without proof.

3.6.3 PROPOSITION

If **X** has a density function f_X and if $y = g(x)$ is a one-to-one differentiable mapping such that $g'(x) \neq 0$ for all x, then $Y = g(X)$ has an absolutely continuous distribution with density function f_Y calculated by

$$f_Y(y) = \frac{f_X(x)}{|g'(x)|}$$

for each y such that $x = g^{-1}(y)$ exists; otherwise $f_Y(y) = 0$.

Example 7 Let $X = (X, Y)$ be a random vector having a density function f_X, and let $Y = (U, V)$ be derived from X as follows:

$$U = aX + bY$$
$$V = cX + dY$$

where the constants a, b, c, d are such that their determinant Δ is not 0. Determine the density function f_Y.

Solution In this example

$$g_1(x, y) = ax + by$$
$$g_2(x, y) = cx + dy$$

and the Borel function $\mathbf{g} = (g_1, g_2)$ is a linear transformation, nonsingular in fact since Δ is assumed nonzero. Thus, $\mathbf{g}$ easily satisfies the condition of Proposition 3.6.3, and we may use the formula to calculate f_Y. Now the Jacobian derivative of $\mathbf{g}$ is none other than the determinant Δ at all $\mathbf{x} = (x, y)$

$$g'(x) = \Delta = ad - bc$$

Also, the inverse transformation $\mathbf{g}^{-1}$ is easily found to be

$$= \frac{d}{\Delta} u - \frac{b}{\Delta} v$$

$$x = -\frac{c}{\Delta} u + \frac{a}{\Delta} v$$

Consequently, we obtain

$$f_{U,V}(u, v) = \frac{f_{X,Y}\left(\dfrac{d}{\Delta} u - \dfrac{b}{\Delta} v, \, -\dfrac{c}{\Delta} u + \dfrac{a}{\Delta} v\right)}{|ad - bc|}$$

In particular, if $b = c = 0$, we have

$$f_{U,V}(u, v) = \frac{f_{X,Y}\left(\dfrac{u}{a}, \dfrac{v}{d}\right)}{|ad|}$$

Example 8 Let $\mathbf{X} = (X, Y)$ have a *bivariate normal density* (cf. Problem 3.6.18) given by

$$f_{X, Y}(x, y) = \frac{1}{2\pi\sqrt{1 - \rho^2}} \exp\left[- \frac{1}{2(1 - \rho^2)} (x^2 - 2\rho xy + y^2)\right]$$

and let $\mathbf{g}$ be a clockwise rotation described by

$$u = \frac{1}{\sqrt{2}} x + \frac{1}{\sqrt{2}} y$$

$$v = - \frac{1}{\sqrt{2}} x + \frac{1}{\sqrt{2}} y$$

If $\mathbf{Y} = (U, V) = \mathbf{g}(\mathbf{X})$, determine the distribution of $\mathbf{Y}$.

Solution We need only apply the result of the preceding example. Here we have $a = b = d = 1/\sqrt{2}$, $c = -1/\sqrt{2}$, and $\Delta = 1$. Therefore

$$f_{U, V}(u, v) = f_{X, Y}\left(\frac{1}{\sqrt{2}} u - \frac{1}{\sqrt{2}} v, \frac{1}{\sqrt{2}} u + \frac{1}{\sqrt{2}} v\right)$$

By direct substitution and simplification we easily obtain

$$f_{U, V}(u, v) = \frac{1}{2\pi\sqrt{1 - \rho^2}} \exp\left\{ - \frac{1}{2(1 - \rho^2)} [(1 - \rho)u^2 + (1 + \rho)v^2]\right\}$$

Since the variables u and v are separated, we can factor $f_{U, V}$ as follows

$$f_{U, V}(u, v)$$

$$= \frac{1}{\sqrt{2\pi}\sqrt{1 + \rho}} \exp\left[- \frac{1}{2}\left(\frac{u}{\sqrt{1 + \rho}}\right)^2\right] \frac{1}{\sqrt{2\pi}\sqrt{1 - \rho}} \exp\left[- \frac{1}{2}\left(\frac{v}{\sqrt{1 - \rho}}\right)^2\right]$$

In other words U and V are independent and, respectively, $(0, \sqrt{1 + \rho})$ and $(0, \sqrt{1 - \rho})$ normally distributed. The significance of this result is that a standardized bivariate normal density $f_{X, Y}$ can be obtained by rotating a product density of two normal densities f_U and f_V.

The condition that $\mathbf{g}$ be one-to-one in Proposition 3.6.3 can be relaxed somewhat to accommodate a wider class of Borel functions, leading to Proposition 3.6.4 stated below. But first we consider the following preliminary definition.

3.6.5 DEFINITION
Given a Borel function $y = \mathbf{g}(x)$ we say that x is a regular point of $\mathbf{g}$ if the Jacobian derivative $\mathbf{g}'(x)$ is nonzero, and that y is a regular point if $\mathbf{g}^{-1}(y)$ consists of countably many regular points $x_1, x_2, x_3, \ldots$.

The following proposition can be regarded as a simple generalization of Proposition 3.6.3.

3.6.4 PROPOSITION

If X has a density function f_X and $y = g(x)$ is such that almost all x are regular in the sense that $P_X(E) = 1$ where E is the set of all regular points of g, then $Y = g(X)$ has an absolutely continuous distribution with density function f_Y calculated by

$$f_Y(y) = \sum_i \frac{f_X(x_i)}{|g'(x_i)|}$$

for each regular point y with $g^{-1}(y) = x_i$. For a nonregular y we simply set $f_Y(y) = 0$.

We illustrate this proposition by a simple example.

Example 9 Let $X = (X, Y)$ be uniformly distributed over the rectangle A with vertexes $(-1, 0)$, $(-1, -1)$, $(1, -1)$, $(1, 0)$, and let the Borel function g be described by

$$u = |x|$$
$$v = -y$$

(See Fig. 3.6.6). If $Y = g(X)$, calculate f_Y.

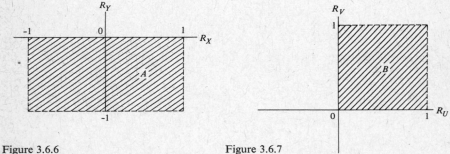

Figure 3.6.6   Figure 3.6.7

Solution Geometrically, what g does is to fold the rectangle A into the square B (see Fig. 3.6.7). Intuitively, Y must have a uniform distribution over B. In terms of Proposition 3.6.4 the only nonregular points in A are those on the R_Y-axis since $g'(x)$ is not even defined there. At regular points in A we have $g'(x) = \pm 1$ depending on whether x is to the left or to the right of R_Y.

In B all the points y are regular except those on R_V-axis. Each regular y has a two-point inverse image $g^{-1}(y) = \{x_1, x_2\}$ in A with $|g'(x_1)| = |g'(x_2)| = 1$. Finally note that $f_X(x) = \frac{1}{2}$ for all x in A. Consequently, by formula, we obtain

$$f_Y(y) = \frac{f_X(x_1)}{|g'(x_1)|} + \frac{f_X(x_2)}{|g'(x_2)|}$$
$$= \tfrac{1}{2} + \tfrac{1}{2} = 1$$

for all regular points y in B.

Case IV

Finally we consider the case $n \neq m \geq 2$. We dismiss the subcase $n < m$ right away since $\mathbf{Y}$ can never be absolutely continuous if $\mathbf{g}$ is a mapping into a higher-dimensional Euclidean space. For example if $n = 1$ and $m = 2$, then the probability mass of $\mathbf{Y}$ will be distributed along a curve in R^2 and consequently $\mathbf{Y}$ cannot possibly have a density function. The subcase $n > m$ is reducible to the case $n = m$ and will be settled in principle by the following proposition.

3.6.5 PROPOSITION

Let $\mathbf{X} = (X, Y, Z)$ be absolutely continuous with density function $f_{\mathbf{X}}(x, y, z)$, and let $\mathbf{g}$ be a Borel function of R^3 into R^2 described by

$$u = g_1(x, y, z)$$
$$v = g_2(x, y, z)$$

If through the intervention of a third Borel function

$$w = g_3(x, y, z)$$

the resulting augmented function $\bar{\mathbf{g}} = (g_1, g_2, g_3)$ is such that

$$\bar{\mathbf{Y}} = \bar{\mathbf{g}}(\mathbf{X})$$

has an absolutely continuous distribution, then $\mathbf{Y} = \mathbf{g}(\mathbf{X})$ has an absolutely continuous distribution with density function obtainable as a marginal density function of $f_{\bar{\mathbf{Y}}}$:

$$f_{\mathbf{Y}}(u, v) = \int_{-\infty}^{\infty} f_{\bar{\mathbf{Y}}}(u, v, w) \, dw$$

We remark that the intervening Borel function g_3 should be chosen as simple as possible. Usually it is simply set as

$$g_3(x, y, z) = z$$

The following example illustrates Proposition 3.6.5 in principle only since we are letting $n = 2$ and $m = 1$.

Example 10 Let $\mathbf{X} = (X, Y)$ be absolutely continuous with density function $f_{\mathbf{X}}$. Determine the density function for the one-dimensional random vector $X + Y$.

Solution We have $\mathbf{g}$ given by

$$u = g_1(x, y) = x + y$$

If we augment $\mathbf{g}$ to $\bar{\mathbf{g}} = (g_1, g_2)$ by

$$u = g_1(x, y) = x + y$$
$$v = g_2(x, y) = y$$

then $\overline{Y} = \overline{g}(X)$ is absolutely continuous (see Proposition 3.6.3 or Example 7 of this section) with density function

$$f_{\overline{Y}}(u, v) = \frac{f_X(u - v, v)}{|\overline{g}'(x, y)|} = f_X(u - v, v)$$

since $x = u - v$, $y = v$, and $\overline{g}'(x, y) = \begin{vmatrix} 1 & 1 \\ 0 & 1 \end{vmatrix} = 1$. Now therefore as a marginal density of $f_{\overline{Y}}$ we obtain

$$f_{X+Y}(u) = f_{\overline{Y}}(u) = \int_{-\infty}^{\infty} f_{\overline{Y}}(u, v) \, dv$$

$$= \int_{-\infty}^{\infty} f_{X,Y}(u - v, v) \, dv$$

3.6 PROBLEMS

1. Let the random vector (X, Y) be uniformly distributed over the circumference of a circle of radius 1 centered at the origin. If $Z = X + Y$, determine $P(Z \leq 1)$. Can you sketch the distribution function F_Z?

2. Let (X, Y) be so distributed that half of the probability mass is distributed uniformly over the square with vertexes $(0, 0)$, $(1, 0)$, $(0, 1)$, $(1, 1)$ and the other half distributed equally among the two points $(0, 1)$ and $(0, 2)$. Describe the distribution of $Z = X + Y$ by sketching the distribution function F_Z.

3. If two independent random variables X and Y are respectively μ and ν Poisson-distributed, show that $X + Y$ is $\mu + \nu$ Poisson-distributed. Hint:

$$\sum_{i=0}^{k} k! / [i!(k - 1)!] \mu^i \nu^{k-1} = (\mu + \nu)^k$$

4. The *generating function* G_X of a random variable X with $\sum_{i=-\infty}^{\infty} P_X(i) = 1$ is defined as follows:

$$G_X(t) = \sum_{i=-\infty}^{\infty} P_X(i) t^i \qquad \text{for} \quad |t| \leq 1$$

a. Show that if X and Y are independent then

$$G_{X+Y}(t) = G_X(t) G_Y(t)$$

b. Use generating functions to show that no matter how two dice are loaded their sums $X + Y = 2, 3, \ldots, 12$ can never be equally likely. Hint: Show by contradiction. Two polynomials, one admitting real roots and the other not, cannot be equal.

5. If X and Y have a joint density function $f_{X,Y}(x, y)$, show that $X + Y$ has a density function given by

$$f_{X+Y}(v) = \int_{-\infty}^{\infty} f_{X,Y}(x, v - x) \, dx$$

Can you interpret this formula geometrically? Sketch $f_{X+Y}(v)$ if X and Y are jointly uniformly distributed over the rectangle with vertexes $(0, 0)$, $(2, 0)$, $(2, 1)$, $(0, 1)$.

6. Optional. Generalize the result of Problem 3.6.5 as follows: If X and Y have a joint density function $f_{X, Y}(x, y)$, then $aX + bY$ has a density function given by

$$f_{aX+bY}(v) = \begin{cases} \dfrac{1}{|b|} \displaystyle\int_{-\infty}^{\infty} f_{X, Y}(x, v - ax/b)\, dx & \text{for} \quad b \neq 0 \\[2ex] \dfrac{1}{|a|} \displaystyle\int_{-\infty}^{\infty} f_{X, Y}(v - by/a, y)\, dt & \text{for} \quad a \neq 0 \end{cases}$$

Note the various special cases by setting a, b to 1, -1, 0, and so on.

7. Optional. If X and Y have a joint density function $f_{X, Y}(x, y)$ show that XY and Y/X have density functions given by

$$f_{XY}(v) = \int_{-\infty}^{\infty} f_{X, Y}(x, v/x)/|x|\, dx$$

$$f_{Y/X}(v) = \int_{-\infty}^{\infty} f_{X, Y}(x, vx) \cdot |x|\, dx$$

We mention a general formula using the gradient:

$$f_{g(X, Y)}(v) = \int_{g^{-1}(v)} f_{X, Y}(x, y)/|\nabla g(x, y)|\, ds$$

where the integral is taken along the curve $g^{-1}(v)$ and ∇g is the gradient vector of the Borel function g.

8. Let X be such that

$$P_X(k) = (\tfrac{1}{2})^k \qquad \text{for} \quad k = 1, 2, 3, \ldots$$

and let Y_n be triangularly distributed over the interval $[-1/n, 1/n]$. Determine the convolution $P_X * P_{Y_n}$. Is $P_X * P_{Y_n}$ continuous for each $n = 1, 2, 3, \ldots$? What happens to $P_X * P_{Y_n}$ when n tends to ∞?

9. Show informally that $P_X * P_Y$ is continuous if and only if one of P_X and P_Y is continuous.

10. Show that if $f_X(x) = \lambda e^{-\lambda x}$ for $x > 0$, then its r-fold convolution is given by

$$(f_X * f_X * \cdots * f_X)(y) = \frac{\lambda^r e^{-\lambda y} y^{r-1}}{(r-1)!} \qquad \text{for} \quad v > 0$$

Hint: Use the result of Example 3.6.6 to calculate $[(f_X * f_X) * f_X](w) = \lambda^3 e^{-\lambda w} w^2/2!$, then proceed inductively.

11. Given a random vector (X, Y) distributed discretely over the points $(1, 0)$, $(0, 1)$, $(1, 1)$ with respective probabilities $\frac{1}{2}$, $\frac{1}{3}$, $\frac{1}{6}$, find a Borel function $g(x, y)$ such that $(U, V) = g(X, Y)$ is distributed discretely over the points $(2, 0)$, $(0, 2)$, $(1, 1)$ with the same respective probabilities $\frac{1}{2}$, $\frac{1}{3}$, $\frac{1}{6}$.

12. Let (X, Y) be distributed uniformly over the unit square $0 \leq x \leq 1$, $0 \leq y \leq 1$, and let (U, V) be such that

$$U = 2X$$
$$V = 3Y$$

Describe the distribution of (U, V).

13. If (X, Y) is uniformly distributed over a disk of radius 1 around the origin, and if (U, V) is defined by

$$U = aX$$
$$V = bY$$

show that (U, V) is uniformly distributed over the ellipse $u^2/a^2 + v^2/b^2 = 1$.

14. Let (X, Y) have a density function $f_{X, Y}(x, y)$, and let (U, V) be defined by

$$U = X + a$$
$$V = Y + b$$

Determine the density function $f_{U, V}(u, v)$.

15. Do the same as in Problem 3.6.14, but with

$$U = cX + a$$
$$V = dY + b$$

where $cd \neq 0$.

16. Let (X, Y) be uniformly distributed over the square with vertexes $(\pm 1, \pm 1)$, and let (U, V) be given by

$$U = X^2$$
$$V = Y^2$$

Show that (U, V) is absolutely continuous with density function

$$f_{U, V}(u, v) = \frac{1}{4\sqrt{uv}}$$

for (u, v) in a certain unit square.

17. Optional. Working as in Example 3.6.10, show that if (X, Y) has a density function $f_{X, Y}(x, y)$ then $X - Y$, XY, and X/Y have density functions as given in Problems 3.6.6 and 3.6.7.

18. Derivation of bivariate normal density function. If U and V are independent and both standard normal, then the joint density function f_0 of U and V is obtained by taking the product of their respective density functions:

$$f_0(u, v) = \frac{1}{2\pi} \exp\{-\tfrac{1}{2}(u^2 + v^2)\}$$

a. Let the Borel function $\mathbf{g}$ of R^2 into R^2 be given by

$$(x', y') = \mathbf{g}(u, v) = (u, v)\begin{bmatrix} 1 & \rho \\ 0 & \sqrt{1 - \rho^2} \end{bmatrix}$$

$$= (u, \rho u + \sqrt{1 - \rho^2}\, v)$$

Show that the density function of $(X', Y') = \mathbf{g}(U, V)$ is given by

$$f_1(x', y') = \frac{1}{2\pi\sqrt{1 - \rho^2}} \exp\left[-\frac{1}{2(1 - \rho^2)}\,(x'^2 - 2\rho x'y' + y'^2)\right]$$

b. Let the Borel function $\mathbf{h}$ of R^2 into R^2 be given by

$$(x'', y'') = \mathbf{h}(x', y') = (x', y')\begin{bmatrix} \sigma_1 & 0 \\ 0 & \sigma_2 \end{bmatrix}$$

$$= (\sigma_1 x', \sigma_2 y')$$

Show that the density function of $(X'', Y'') = \mathbf{h}(X', Y')$ is given by

$$f_2(x'', y'') = \frac{1}{2\pi\sigma_1\sigma_2\sqrt{1 - \rho^2}} \exp\left\{-\frac{1}{2(1 - \rho^2)}\left[\left(\frac{x''}{\sigma_1}\right)^2 - 2\rho\left(\frac{x''}{\sigma_1}\right)\left(\frac{y''}{\sigma_2}\right) + \left(\frac{y''}{\sigma_2}\right)^2\right]\right\}$$

c. Let the Borel function $\mathbf{k}$ of R^2 into R^2 be given by

$$(x, y) = \mathbf{k}(x'', y'') = (x'', y'') + (\mu_1, \mu_2)$$

$$= (x'' + \mu_1, y'' + \mu_2)$$

Show that the density function of $(X, Y) = \mathbf{k}(X'', Y'')$ is given by

$$f_{X, Y}(x, y) = \frac{1}{2\pi\sigma_1\sigma_2\sqrt{1 - \rho^2}} \exp\left\{-\frac{1}{2(1 - \rho^2)}\left[\left(\frac{x - \mu_1}{\sigma_1}\right)^2\right.\right.$$

$$\left.\left. - 2\rho\left(\frac{x - \mu_1}{\sigma_1}\right)\left(\frac{y - \mu_2}{\sigma_2}\right) + \left(\frac{y - \mu_2}{\sigma_2}\right)^2\right]\right\}$$

This last function is known as a $(\mu_1, \mu_2, \sigma_1, \sigma_2, \rho)$ *bivariate normal density function.*

Chapter 4
MATHEMATICAL
EXPECTATIONS

The basic idea of mathematical expectation is to find an average value for a given random variable. In Section 4.1 (Expected Value of a Random Variable), we indicate how this is done, separately for discrete random variables and continuous random variables having density functions. The two seemingly unrelated methods of finding the average (or expected) value are then shown to be actually two consequent formulas of one unifying definition of mathematical expectation. Such a unifying definition satisfies an aesthetic requirement that one and only one mathematical definition should be given to each mathematical concept. Aside from this, the definition also introduces in a natural context the Lebesgue way of integration as against the more traditional Riemann way of integration. The essential difference between these two approaches to integration is explained at the end of Section 4.1. If one is only interested in actual calculations of expected values, the latter part of Section 4.1 as well as its continuation, Section 4.2 (Mathematical Expectation as an Operator), may be omitted.

In Section 4.3 (Numerical Characterizations of a Random Variable), we apply our formulas of mathematical expectation to a random variable in more than one way to extract the average value (i.e., the *mean*), the average deviation from the average value (i.e., the *standard deviation*), and so on. These numerical characterizations provide a crude profile for the probability distribution of a random variable; in many cases, they turn out to be all one needs in deducing important results about a random variable. Of course, the numerical characterizations considered in Section 4.3 fall far short by themselves of completely characterizing the distribution of a random variable; extensions of these characterizations lead eventually to the moment-generating function and the characteristic function, which completely characterizes the distribution of a random variable (cf. Section 6.3).

Section 4.4 (Numerical Characterizations of a Random Vector) parallels Section 4.3 in an obvious way. However, we obtain here two types of numerical characterizations: the first type pertains to individual component random variables, while the second type pertains to two or more random variables simultaneously. Although the second type is the more important, we cannot dwell on it without having recourse to *conditional distributions and expectations*. These, however, will not be considered until the next chapter.

4.1 EXPECTED VALUE OF A RANDOM VARIABLE

Although a random variable may assume different values in a wide range, in some sense there must be some sort of an average that summarizes in one single number our "expectation" of this random variable as a whole. For example, if X denotes the number of dots on the upturned face of an ordinary die, the "expected value" of X should be given by

$$E(X) = \frac{1 + 2 + 3 + 4 + 5 + 6}{6} = 3.5$$

Now, of course, this does not mean that every time we throw an ordinary die we expect to see 3.5 dots; this means rather that the number of dots we see will be in some not too large neighborhood of 3.5 because these numbers average out to 3.5.

Rewriting $E(X)$ above as

$$E(X) = 1(\tfrac{1}{6}) + 2(\tfrac{1}{6}) + 3(\tfrac{1}{6}) + 4(\tfrac{1}{6}) + 5(\tfrac{1}{6}) + 6(\tfrac{1}{6})$$

we regard it as a probability-weighted sum of the values of X, or "probability-weighted average" of X. This suggests that for a loaded die, the expected value of X will be given by

$$E(X) = 1 \cdot p_1 + 2 \cdot p_2 + \cdots + 6 \cdot p_6$$

where p_i denotes the probability that X assumes the value i. If for example the die is so extremely loaded that $p_1 = 1$, then $E(X) = 1$, or if it is so loaded that p_6 is nearly 1, then $E(X)$ is nearly 6. If this die is thrown a large number of times, say, N, it will give 1 roughly Np_1 times, 2 roughly Np_2 times, and so on, so that the average number of dots in the long run will be roughly

$$\frac{1(Np_1) + 2(Np_2) + \cdots + 6(Np_6)}{N}$$

which, after cancellations of N, is the $E(X)$ just considered above. We can now consider the following more general definition.

4.1.1 DEFINITION

Given a discrete random variable X which takes the values $x_1, x_2, x_3, \ldots$ with the corresponding probabilities $p_1, p_2, p_3, \ldots$, we define its *expected value* by

$$E(X) = x_1 p_1 + x_2 p_2 + \cdots = \sum_i x_i p_i$$

Example 1 An ordinary coin is tossed till a head appears. How many times, on the average, must one toss the coin?

Solution Denoting by X the number of tosses till a head appears, we calculate the expected value of X as follows

$$E(X) = 1p_1 + 2p_2 + 3p_3 + \cdots$$

where $p_i = P_X(i) = (\tfrac{1}{2})^i$ so that

$$E(X) = \sum_{k=1}^{\infty} k(\tfrac{1}{2})^k$$

Now the sum of this series can be found from the sum of geometric series

$$\sum_{k=0}^{\infty} x^k = \frac{1}{1-x}$$

Differentiating this series with respect to x, we obtain

$$\sum_{k=1}^{\infty} k x^{k-1} = \frac{1}{(1-x)^2}$$

Consequently we have

$$E(X) = \frac{1}{2} \sum_{k=1}^{\infty} k(\tfrac{1}{2})^{k-1} = \frac{1}{2} \frac{1}{(1-\tfrac{1}{2})^2} = 2$$

Although we set $E(X) = \sum_i x_i p_i$ in Definition 4.1.1, we must give up defining $E(X)$ if X is such that both the positive and the negative parts of $\sum_i x_i p_i$ diverge to ∞ (see Problem 4.1.2 for such an X). In this case, $\sum_i x_i p_i$ does not have a meaningful sum, and we shall call such a series *nonsummable*.

Definition 1 applies only to discrete X. What if X takes more than just countably many different values? First, we shall consider $E(X)$ for an absolutely continuous X leading to Definition 4.1.2, then we shall develop a definition of $E(X)$ consistent with Definitions 4.1.1–2 that applies to all X regardless of types of distribution.

If X has a density function f_X, it takes values in a small interval $\Delta_i x$ with probability approximately equal to $f_X(x_i) \Delta_i x$ where x_i is some point in the interval $\Delta_i x$. Now although there are numerous values that X can take in the interval $\Delta_i x$, these values are all approximately equal to x_i. Therefore, approximating X with the discrete random variable that takes the values x_i with the corresponding probabilities $p_i = f_X(x_i) \, \Delta_i x$, we see that the expected value of this discrete random variable

$$\sum_i x_i p_i = \sum_i x_i f_X(x_i) \, \Delta_i x$$

should be close to what we want to call $E(X)$ for the original X. Now, if as we let all the interval lengths $|\Delta_i x|$ shrink to 0 the above approximating sum tends to a limit, then we should take this limit as the expected value for X. But the limit, if it exists, is exactly the Riemann integral $\int_{-\infty}^{\infty} x f_X(x) \, dx$. Consequently we are led to the following definition.

4.1.2 DEFINITION

Given a continuous random variable X having a density function f_X, we define its expected value by

$$E(X) = \int_{-\infty}^{\infty} x f_X(x) \, dx$$

Here again, $E(X)$ is not defined when both the positive and the negative parts of the integral are infinite.

Example 2 A real number X is drawn at random from the interval $[a, b]$. Calculate the expected value $E(X)$.

Solution Intuitively, the average value of X is $\frac{1}{2}(a + b)$. Or, since the probability mass of X is uniformly spread over the interval $[a, b]$, the center of mass at the midpoint of $[a, b]$ should be taken as the expected value of X. Formally, we calculate $E(X)$ by using Definition 4.1.2. From

$$f_X(x) = \frac{1}{b - a} \qquad \text{for} \quad x \text{ in } [a, b]$$

we obtain easily

$$E(X) = \int_a^b x \, \frac{1}{b - a} \, dx = \tfrac{1}{2}(a + b)$$

We now proceed to develop a general definition of $E(X)$.

Given a random variable X defined on a probability space $(S, \mathscr{S}, P)$, let the range R be partitioned into intervals $\Delta_j x$ by means of integer points $0, \pm 1, \pm 2, \ldots$ (see Fig. 4.1.1), and let $X^{-1}(\Delta_j x)$ be denoted by $\Delta_j \omega$. Then

$$\Delta_j \omega = X^{-1}(\Delta_j x) \qquad j = \pm 1, \pm 2, \ldots$$

is an event consisting of outcomes ω that are assigned values in $\Delta_j x$ by X. If we let x_j^* be the midpoint in $\Delta_j x$, then roughly

$$X(\Delta_j \omega) \doteq x_j^*$$

meaning that X takes each ω in $\Delta_j \omega$ to a point close to x_j^*. Consequently the sum

$$\sigma_0 = \sum_j x_j^* P(\Delta_j \omega) \qquad \text{(assume summability)}$$

should be close to whatever we want to define as the expected value of X.

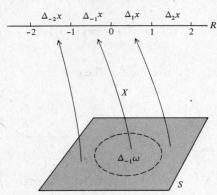

Figure 4.1.1

In other words, σ_0 is a sum approximating what we are trying to reach. However, σ_0 can be easily replaced by an even better approximating sum. In fact, if we refine our partition of R by bisecting each existing interval and then proceed as before, we obtain a "better" approximating sum σ_1. More generally, let R be partitioned into intervals $\Delta_j x$ by means of division points 0, $\pm 1/(2^N)$, $\pm 2/(2^N)$, $\pm 3/(2^N)$, ..., where N is a positive integer, and let x_j^* be the midpoint of the interval $\Delta_j x$

$$\Delta_j x = \left(\frac{j-1}{2^N}, \frac{j}{2^N} \right)$$

Then the sum

$$\sigma_N = \sum_j x_j^* P(\Delta_j \omega)$$

should approach what we want to call the expected value of X as N approaches ∞. Therefore, if indeed $\lim_{N \to \infty} \sigma_N$ exists (presupposing the summability for each σ_N). then this limit ought to be defined as the expected value of X. Accordingly we now formally state the following definition.

4.1.3 DEFINITION

Given a random variable X on a probability space $(S, \mathscr{S}, P)$, the expected value of X is defined, whenever possible, by

$$E_S(X) = \lim_{N \to \infty} \sigma_N = \lim_{N \to \infty} \sum_j x_j^* P(\Delta_j \omega)$$

where the σ_N are the approximating sums described just previously.

It is not difficult to see that this definition of expected value of X is consistent with the earlier one for discrete X. Consider for example a discrete X having a two-point distribution $P_X(0) = P_X(1) = \frac{1}{2}$. Applying the limiting process above, we can see that

$$E_S(X) = \tfrac{1}{2} = E(X)$$

It is more difficult to see that our general definition of $E_S(X)$ also agrees with the earlier one for absolutely continuous X. However, this will be shown in Corollary 4.2.1 of the next section.

The elaborate mathematical process applied to X to arrive at the limiting value $E_S(X)$ deserves a special name and a notation summarizing the entire process. Since this process is originally due to Lebesgue, and is "dual" in a sense to Riemann's integration process, we shall call it the *Lebesgue abstract integration*, and imitating the notation for the Riemann integration, we shall denote it by

$$\int_S X(\omega)P(d\omega) \quad \text{or briefly} \quad \int_S X \, dP$$

This integral notation says that a random variable (or $\mathscr{S}$-measurable function) X is being Lebesgue-integrated on the probability space (or measure space) S with respect to the probability measure (or measure) P.

Using the notation of Lebesgue abstract integral, we may rephrase our Definition 4.1.3 as follows.

4.1.4 DEFINITION

Given a random variable X on a probability space $(S, \mathscr{S}, P)$, its expected value is defined by

$$E_S(X) = \int_S X(\omega)P(d\omega)$$

whenever the integral exists.

The main difference between the Riemann way and the Lebesgue way of integration lies in the fact that the approximating sums are formed in the former case by subdivision of the domain, and in the latter case by subdivision of the range, of the integrand function (or random variable in our case). In the former case, it is naively assumed that the function values remain more or less unchanged within each subdivided domain (or subinterval). This means that a Riemann approximating sum is not adequate when the function being integrated is very "jumpy" or "highly" discontinuous. However, the Riemann integral has been so carefully defined that whenever the approximating sums are inadequate, the function is declared (Riemann) unintegrable rather than assigned an inadequate integral value. In contrast, Lebesgue approximating sums are adequate whenever they are formed, and the limiting value of these approximating sums is defined as the (Lebesgue) integral of the function.

If the Riemann integral is defined for a certain (reasonably continuous) function, the integral value assigned is no different from that assigned by the Lebesgue integral. On the other hand, the Riemann integral may not be defined for a
certain (highly discontinuous) function, and yet the Lebesgue integral may be
defined for this particular function. Finally, even the Lebesgue integral may fail
if the function does not permit approximating sums to be formed. For example,
if X on $(S, \mathscr{S}, P)$ is *not* a random variable, then for a certain $\Delta_j x$ the corresponding $\Delta_j \omega = X^{-1}(\Delta_j x)$ may not belong to $\mathscr{S}$ so that $P(\Delta_j \omega)$ is not defined and
consequently neither is the approximating sum $\sum_j x_j^* P(\Delta_j \omega)$.

We have used the term Lebesgue *abstract* integral in this section because the
domain of integration is a general probability space, which lacks *concrete*
mathematical structure such as found in the Euclidean spaces. In the next section
we shall consider random variables defined on Euclidean probability spaces and
arrive at an important special case of the Lebesgue abstract integral known as
Lebesgue-Stieltjes integral. However, owing to technical complexities inherent
in the subject, some results are given without proofs (cf. Minibibliography:
Royden). The next section may be omitted, provided that problems of the present
section, especially those of a general nature, are paid due attention.

4.1 PROBLEMS

1. A coin with probability p of falling heads is tossed till it falls heads. How many
times on the average must one toss the coin?

2. A gambler tosses a fair coin till it falls heads. If the coin is tossed k times altogether,
he wins or loses $2^k/k$ dollars according as k is odd or even. What is his expected gain?

3. If a discrete random variable X is squared so that $Y = X^2$, can the expected value
of Y be obtained by squaring that of X, that is, does $E(X^2) = [E(X)]^2$? Hint: Let X
be the number of heads in a single toss of a coin.

4. If a discrete random variable X is doubled so that $Y = 2X$, can the expected value
of Y be obtained by doubling that of X, that is, does $E(2X) = 2E(X)$?

5. If a discrete random variable X is augmented by a constant a so that $Y = X + a$,
can the expected value of Y be obtained from that of X by adding a, that is, does
$E(X + a) = E(X) + a$?

6. If X has a discrete distribution, so does $g(X)$ for any Borel function g. With the
formula in Definition 4.1.1 in view, derive somewhat intuitively the formula

$$E[g(X)] = \sum_i g(x_i)p_i$$

where $P_i = P(X = x_i)$. Use this formula to calculate $E(X^2 + 1)$ if $P(X = -1) = \frac{1}{4}$,
$P(X = 1) = \frac{1}{4}$, and $P(X = 2) = \frac{1}{2}$.

7. If X has an absolutely continuous distribution, so does $g(X)$ for some well-behaved g. With the formula in Definition 4.1.2 in view, derive somewhat intuitively the formula

$$E[g(X)] = \int_{-\infty}^{\infty} g(x)f_X(x)\, dx$$

where f_X is a density function of X. Use this formula to calculate $E(X^2)$ if X has a uniform distribution over the interval $[0, 1]$.

8. Three balanced dice are tossed. You can either bet on a triple, (e.g., three 5s), in which event you will be awarded a number of dollars equal to the total number of dots, or you can bet on the ace, in which event you will be awarded 70¢ for each ace that appears. From the point of view of mathematical expectation, which bet should you choose?

9. An urn contains 5 balls labeled 1 to 5. Three balls are drawn simultaneously from the urn. Let X be the largest number observed. Calculate the expected value of X.

10 a. A man pays \$1 to buy a lottery ticket in a lottery that sells a total of 10,000 tickets and that gives 1 grand prize of \$5000, 2 second prizes of \$1000 each, and 30 third prizes of \$100 each. Let X be the man's net gain and calculate $E(X)$. If the price of the ticket is \$1.05, would it necessarily be foolish to buy such a ticket?

    b. In a casino a fair coin is tossed till it falls heads up to a maximum of 100 tosses. If it falls heads (for the first time) at the nth toss you are paid $X = 2^n$ dollars. Find $E(X)$. Would you be willing to pay the entrance fee equal to $E(X)$ in order to take part in the gamble? Can you give a probabilistic reason for your negative decision? (See Minibibliography: Feller, on Petersburg paradox.)

11. A roulette wheel has 37 pockets labeled from 0 to 36. A ball tossed as the wheel spins eventually settles in one of these pockets. Two players A and B each having N dollars at his disposal play roulette as follows: If the outcome is an odd number, A takes a dollar from B, otherwise B takes a dollar from A. B is obliged to continue the game as long as neither is broke and A insists on playing. A, however, may terminate the game anytime he wishes. Which player would you rather be? Can you give a reasonable explanation for your choice? If $N = 2$, what would be your choice? What if $N = 1$?

12. When we say that X has a discrete distribution with $\{x_1, x_2, x_3, \ldots\}$ and $\{p_1, p_2, p_3, \ldots\}$ so that $E(X) = \sum_i x_i p_i$, $x_1, x_2, x_3, \ldots$ are understood to be distinct points, and $p_1, p_2, p_3, \ldots$ are understood to be positive real numbers totaling 1. However, a discrete distribution may also be described by a sequence of not necessarily distinct points $\{z_1, z_2, z_3, \ldots\}$ and an accompanying sequence of nonnegative real numbers totaling 1, say, $\{q_1, q_2, q_3, \ldots\}$. In this situation we have obviously $E(X) = \sum_i x_i p_i = \sum_j z_j q_j$. With this in mind, show that if $Z = g(X, Y)$ and (X, Y) has a discrete distribution with $\{(x_1, y_1), (x_2, y_2), \ldots\}$ and $\{p_1, p_2, \ldots\}$, then

$$E(Z) = E[g(X, Y)] = \sum_i g(x_i, y_i)p_i$$

13 a. With the formula in the preceding problem show that if X and Y have a discrete joint distribution, then

$$E(aX + bY) = aE(X) + bE(Y)$$

for any real numbers a and b.

b. Find a discrete joint distribution of X and Y for which the following formula does not hold:

$$E(XY) = E(X)E(Y)$$

c. If (X, Y) has a discrete distribution, and X and Y are independent, show that $E(XY) = E(X)E(Y)$.

14. A coin with probability p of falling heads is tossed n times. If X denotes the number of heads in these n tosses, and X_i denotes the number (either 0 or 1) of heads occurring at the ith toss, then $X = X_1 + X_2 + \cdots + X_n$. Use the formula in Problem 4.1.13(a) to show that $E(X) = np$.

15 a. If (X, Y) has a continuous distribution with a density function $f_{X,\,Y}(x, y)$ and $g(x, y)$ is a well-behaved function, derive somewhat intuitively the formula

$$E[g(X, Y)] = \int_{-\infty}^{\infty} \int_{-\infty}^{\infty} g(x, y) f_{X,\,Y}(x, y)\, dy\, dx$$

b. With this formula show that if X and Y have an absolutely continuous joint distribution, then

$$E(aX + bY) = aE(X) + bE(Y)$$

and further, if X and Y are independent, then

$$E(XY) = E(X)E(Y)$$

4.2 MATHEMATICAL EXPECTATION AS AN OPERATOR

For any random variable X on a probability space $(S, \mathscr{S}, P)$ we have defined the expected value $E_S(X)$. Clearly, as X is varied, $E_S(X)$ will also vary; thus E_S can be regarded as a "function" (or mapping) assigning to each random variable X on S a unique real number $E_S(X)$. The mapping E_S has as its domain the class of all random variables (for which the expected values exist) defined on S, and as its range the real numbers. Since random variables are essentially functions themselves, following the general usage of the word we shall call E_S an *operator*. E_S "operates" on X to produce a number $E_S(X)$. The operator E_S (also denoted E when there is no fear of confusion) will be called the *mathematical expectation*. The purpose of this section is to study the properties of the mathematical expectation E_S.

What kind of an operator is E_S? Among other things we shall see that E_S is a "linear" operator in the sense that

$$E_S(aX + bY) = aE_S(X) + bE_S(Y)$$

Such an observation is useful in calculating the expected value of a random variable which is expressible as a linear combination of simpler random variables.

Since E_S was defined in terms of Lebesgue abstract integration, the investigation of E_S naturally falls back on the study of the properties of the Lebesgue abstract integral. However, as far as we are concerned, it suffices to merely note that in abstract integration theory it is shown quite routinely that most of the familiar properties of the Riemann integral are also valid for the Lebesgue abstract integral. Specifically, we have

(i) $\int_S X(\omega)\, P(d\omega) \geq 0$ if $X(\omega) \geq 0$ for all ω in S.

(ii) $\int_S I_A(\omega)\, P(d\omega) = P(A)$ where I_A is the *indicator function* of A sending $A \subset S$ to 1 and the rest $(S - A)$ to 0.

(iii) $\int_S (aX + bY)(\omega)\, P(d\omega) = a \int_S X(\omega)\, P(d\omega) + b \int_S Y(\omega)\, P(d\omega)$.

Therefore in terms of E_S we correspondingly have

(i) $E_S(X) \geq 0$ if $P(X \geq 0) = 1$.

(ii) $E_S(I_A) = P(A)$.

(iii) $E_S(aX + bY) = aE_S(X) + bE_S(Y)$.

We illustrate the application of (iii) by the following example.

Example 1 Calculate $E(X)$ if X denotes the number of heads in n tosses of a p-coin.

Solution Let X_i be the random variable indicating the number of heads for the ith toss only. Thus $P(X_i = 1) = p$ and $P(X_i = 0) = q = 1 - p$. Now clearly

$$X = X_1 + X_2 + \cdots + X_n$$

Hence by the linearity of E, we obtain

$$E(X) = E(X_1) + E(X_2) + \cdots + E(X_n)$$

but trivially $E(X_i) = 1 \cdot p + 0 \cdot q = p$ for $i = 1, 2, \ldots, n$. Consequently

$$E(X) = np$$

In addition to the three basic properties above, we now derive an extremely useful property of E. This property is essential for actual calculations of $E(X)$; it also is the key to seeing the consistency of Definitions 4.1.3 and 4.1.2.

Consider the following diagram

$$(T, \mathscr{T}, Q)$$

$$(S, \mathscr{S}, P) \xrightarrow[\phi(\xi)]{} R$$

in which ξ is a mapping of a probability space S into a probability space T such that

(a) $$\xi^{-1}(B) \in \mathscr{S} \quad \text{for any} \quad B \in \mathscr{T}$$

(b) $$Q(B) = P[\xi^{-1}(B)]$$

and ϕ is a random variable on the probability space T, that is,

$$\phi^{-1}(B) \in \mathscr{T} \quad \text{for any Borel set} \quad B \text{ in } R$$

Such a setup is not entirely new. For example, as a special case of the diagram given at the beginning of Section 3.6, we have the following diagram.

$$(R^2, \mathscr{B}^2, P_{X,Y})$$

$$(S, \mathscr{S}, P) \xrightarrow[g(X,Y)]{} R$$

where (X, Y) is a random vector (hence an instance of ξ stipulated above), and g is a Borel function (hence an instance of ϕ).

Now the composition of ξ and ϕ defined by

$$[\phi(\xi)](\omega) = \phi[\xi(\omega)]$$

is a random variable on S, hence it makes sense to consider $E_S[\phi(\xi)]$. But ϕ is also a random variable though on a different probability space T, hence it makes sense to consider $E_T(\phi)$. Now what is of great importance for us here is that

(iv) $E_S[\phi(\xi)] = E_T(\phi)$

so that if it is difficult to evaluate $E_S[\phi(\xi)]$ we can always try $E_T(\phi)$, or vice versa.

How do we see that (iv) holds? Consider Fig. 4.2.1. According to Definition 4.1.3, $E_S[\phi(\xi)]$ is the limit of the approximating sum

$$\sum_j x_j^* P(\Delta_j \omega)$$

where $\Delta_j \omega = [\phi(\xi)]^{-1}(\Delta_j x)$ and, as we recall, x_j^* is the midpoint of the interval $\Delta_j x$, while $E_T(\phi)$ is the limit of the approximating sum

$$\sum_j x_j^* Q(\Delta_j \lambda)$$

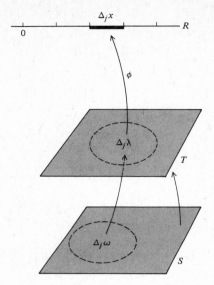

Figure 4.2.1

where $\Delta_j \lambda = \phi^{-1}(\Delta_j x)$. But these two approximating sums are actually equal since

$$\Delta_j \omega = [\phi(\xi)]^{-1}(\Delta_j x) = \xi^{-1}[\phi^{-1}(\Delta_j x)] = \xi^{-1}(\Delta_j \lambda)$$

and by condition (b) of ξ

$$P(\Delta_j \omega) = P[\xi^{-1}(\Delta_j \lambda)] = Q(\Delta_j \lambda)$$

Consequently,

$$E_S[\phi(\xi)] = E_T(\phi)$$

In terms of the Lebesgue abstract integral we may state this result as the following proposition.

4.2.1 PROPOSITION

Given a *measure-preserving* mapping ξ from $(S, \mathscr{S}, P)$ to $(T, \mathscr{T}, Q)$, that is, $Q(B) = P[\xi^{-1}(B)]$ for all B in $\mathscr{T}$, and a random variable ϕ on T, the Lebesgue abstract integral of the random variable $\phi(\xi)$ on S is equal to the Lebesgue abstract integral of the random variable ϕ on T,

$$\int_S [\phi(\xi)](\omega)\, P(d\omega) = \int_T \phi(\lambda)\, Q(d\lambda)$$

This equation becomes especially useful when $(T, \mathscr{T}, Q)$ happens to be an Euclidean probability space $(R^n, \mathscr{B}^n, Q)$ for $n = 1, 2, 3, \ldots$. We elaborate on this by stating the following definition and propositions.

4.2.1 DEFINITION

A Lebesgue abstract integral taken over an Euclidean probability (or measure) space $(R^n, \mathscr{B}^n, Q)$ is called a Lebesgue-Stieltjes integral. The notations for such integrals are

$$\int_{R^n} \phi(x)\, Q(dx) \qquad \text{for} \quad n \geq 2$$

and

$$\int_{R} \phi(x)\, Q(dx) \qquad \text{for} \quad n = 1$$

4.2.2 PROPOSITION

The Lebesgue-Stieltjes integral of $\phi(x)$ on $(R^n, \mathscr{B}^n, Q)$ is equal to the Riemann-Stieltjes integral of $\phi(x)$ if ϕ is continuous everywhere except at some isolated points.

A proof of this proposition belongs to integration theory. Here we will merely point out that a Riemann-Stieltjes integral is the limit of the approximating sums arising from partitions of the domain of ϕ instead of the range of ϕ as was the case with the Lebesgue-Stieltjes integral, and that for a "sufficiently" continuous function ϕ it does not really matter whether we partition the range or the domain of ϕ in forming the approximating sums to arrive at the limit.

4.2.3 PROPOSITION

The Riemann-Stieltjes integral of $\phi(x)$ on $(R^n, \mathscr{B}^n, Q)$ is equal to an ordinary Riemann integral if Q has a density function $f(x)$. Specifically,

$$\int_{R^n} \phi(x)\, Q(dx) = \int_{R^n} \phi(x) f(x)\, dx$$

We can appreciate the plausibility of this equation by looking at the corresponding approximating sums:

$$\sum_i \phi(x_i) Q\,(\Delta_i x) \doteq \sum_i \phi(x_i) f(x_i)\, \Delta_i x$$

Now we are ready for the following corollary of all three preceding propositions, which in effect identifies Definitions 4.1.2 and 4.1.3 of mathematical expectations for absolutely continuous random variables.

4.2.1 COROLLARY

If X defined on $(S, \mathscr{S}, P)$ is absolutely continuous with a Riemann integrable density function f_X, then

$$E_S(X) = \int_{-\infty}^{\infty} x f_X(x)\, dx$$

PROOF
Consider the following diagram

$$(R, \mathscr{B}, P_X)$$

$$X \nearrow \qquad \searrow e$$

$$(S, \mathscr{S}, P) \xrightarrow[\;e(X) = X\;]{} R$$

where e is the identity function, $e(x) = x$ for all x in R. By Proposition 4.2.1 we have

$$E_S[e(X)] = \int_S [e(X)](\omega)\, P(d\omega) = \int_R e(x)\, P_X(dx) \qquad \text{[Lebesgue-Stieltjes]}$$

But, by Proposition 4.2.2 the above Lebesgue-Stieltjes integral is equal to the corresponding Riemann-Stieltjes integral, which by Proposition 4.2.3 is equal to an ordinary Riemann integral, namely,

$$\int_R e(x)\, P_X(dx)\text{[Riemann-Stieltjes]} = \int_{-\infty}^{\infty} x f_X(x)\, dx$$

Consequently,

$$E_S(X) = E_S[e(X)] = \int_{-\infty}^{\infty} x f_X(x)\, dx$$

The following corollary is in a similar vein to that of Corollary 4.2.1, but has a much wider application.

4.2.2 COROLLARY
If X is an n-dimensional random vector defined on $(S, \mathscr{S}, P)$, and g is a real-valued Borel function of an n-dimensional vector variable, then

$$E_S[g(\mathbf{X})] = \int_{R^n} g(x)\, P_{\mathbf{X}}(dx) \qquad \text{[Lebesgue-Stieltjes]}$$

In particular, if g is continuous and $\mathbf{X}$ has a Riemann integrable density function $f_{\mathbf{X}}$, then

$$E_S[g(\mathbf{X})] = \int_{R^n} g(x) f_{\mathbf{X}}(x)\, dx \qquad \text{[Riemann integral]}$$

PROOF
Consider the following diagram

$$(R^n, \mathscr{B}^n, P_{\mathbf{X}})$$

$$\mathbf{X} \nearrow \qquad \searrow g$$

$$(S, \mathscr{S}, P) \xrightarrow[\;g(X)\;]{} R$$

Now by Proposition 4.2.1 we have

$$E_S[g(\mathbf{X})] = E_{R^n}(g) = \int_{R^n} g(x)\, P_{\mathbf{X}}(dx)$$

If g is sufficiently continuous, and $\mathbf{X}$ has a Riemann integrable density function $f_{\mathbf{X}}$, then by Propositions 4.2.2 and 4.2.3 we have

$$\int_{R^n} g(x)\, P_{\mathbf{X}}(dx) = \int_{R^n} g(x) f_{\mathbf{X}}(x)\, dx$$

and the proof is complete.

Note that Corollary 4.2.2 actually includes Corollary 4.2.1 as a special case. To see this we need only let $n = 1$ and set $g = e$ (the identity function).

Example 2 Let X be a random variable having a uniform distribution over the interval $[0, 1]$. We already know $E(X) = \frac{1}{2}$ from Example 2 of Section 4.1. Now calculate $E(X^2)$.

Solution Letting $g(x) = x^2$ in Corollary 4.2.2, we obtain, since $f_X(x) = 1$,

$$E(X^2) = \int_0^1 x^2 \cdot 1 \cdot dx = \left.\frac{x^3}{3}\right|_0^1 = \frac{1}{3}$$

Example 3 Let (X, Y) have a joint density function $f_{X,Y}\,(x, y)$. Calculate the "marginal" expected values for X and Y in terms of the joint density function of X and Y.

Solution Consider the following diagram

$$\begin{array}{ccc}
& (R^2, \mathscr{B}^2, P_{X,Y}) & \\
{\scriptstyle (X,\,Y)}\nearrow & & \searrow {\scriptstyle j} \\
(S, \mathscr{S}, P) & \xrightarrow[\;\;j(X,\,Y)\,=\,X\;\;]{} & R
\end{array}$$

where j is the projection function $j(x, y) = x$ for all (x, y) in R^2. By Corollary 4.2.2, we have

$$E(X) = E_S[j(X, Y)] = E_{R^2}(j) = \int_{R^2} j(x, y) f_{X,Y}(x, y)\, d(x, y)$$

$$= \int_{-\infty}^{\infty} \int_{-\infty}^{\infty} x f_{X,Y}(x, y)\, dy\, dx$$

Likewise we obtain

$$E(Y) = \int_{-\infty}^{\infty} \int_{-\infty}^{\infty} y f_{X,Y}(x, y) \, dy \, dx$$

We conclude this section by discussing one last basic property of the mathematical expectation E_S. Considering the fact that the third property of E_S essentially says the expected value of a sum is the sum of the expected values, we naturally wonder whether the expected value of a product is likewise the product of the expected values. That is, given X and Y on $(S, \mathscr{S}, P)$ can we say in general that

$$E_S(XY) = E_S(X)E_S(Y)$$

The answer is no, as illustrated by the following simple example.

Example 4 A coin is tossed. Let X and Y denote the number of heads and the number of tails for each toss. Calculate $E(X)$, $E(Y)$, $E(XY)$, and compare the latter with $E(X)E(Y)$.

Solution Assuming the coin falls heads with probability p, we have

$$E(X) = 1 \cdot p + 0 \cdot q = p$$
$$E(Y) = 1 \cdot q + 0 \cdot p = q$$

On the other hand XY is constantly 0, so that $E(XY) = 0$. Hence

$$E(XY) \neq E(X)E(Y)$$

Now that we know that the expected value of the product is in general not equal to the product of the expected values, we may ask under what condition the equality holds. The answer is formulated as the fifth property of E_S.

(v) $E_S(XY) = E_S(X)E_S(Y)$ if X and Y are independent

A complete justification of (v) requires Fubini's theorem on integration over a product measure space (cf. Minibibliography: Royden). But the gist of it is as follows. Consider the diagram

$$(R^2, \mathscr{B}^2, P_{X,Y})$$

$$(X, Y) \nearrow \qquad \searrow \phi$$

$$(S, \mathscr{S}, P) \xrightarrow[\phi(X, Y) = XY]{} R$$

in which ϕ is the product function $\phi(x, y) = xy$ for all (x, y) in R^2. Now according to Proposition 4.2.1:

$$E_S(XY) = E_{R^2}(\phi) = \int_{R^2} xy \, P_{X,Y}(d(x, y))$$

If X and Y are independent, then $P_{X,Y}$ is merely the product measure $P_X \times P_Y$. Now according to Fubini, an integration over a product measure space can be evaluated by iterated integration over component measure space. Specifically,

$$\int_{R^2} \phi(x, y)(P_X \times P_Y)(d(x, y)) = \int_R \left[\int_R \phi(x, y) P_Y(dy) \right] P_X(dx)$$

In particular, for $\phi(x, y) = xy$, we have

$$\int_{R^2} xy \,(P_X \times P_Y)(d(x, y)) = \int_R \left[\int_R xy \, P_Y(dy) \right] P_X(dx)$$

$$= \int_R \left[x \int_R y \, P_Y(dy) \right] P_X(dx)$$

$$= \left[\int_R x \, P_X(dx) \right] \left[\int_R y \, P_Y(dy) \right]$$

$$= E_S(X)E_S(Y)$$

In Problem 4.2.4 the reader is asked to prove (v) under a special assumption.

4.2 PROBLEMS

1. Show that if X and Y have a Riemann integrable joint density function, then

$$E(X+Y) = E(X) + E(Y)$$

Hint: In Section 4.2, Corollary 2, let $g(x, y) = x + y$ and use the additivity property of the ordinary Riemann integral.

2. Show that if X and Y have a discrete joint distribution, then

$$E(X+Y) = E(X) + E(Y)$$

Hint: Again use Section 4.2, Corollary 2, plus the fact that for a discrete P_X the Lebesgue-Stieltjes integral $\int_{R^n} g(\mathbf{x}) P_X(d\mathbf{x})$ reduces to $\sum_i g(\mathbf{x}_i) P_X(d\mathbf{x})$ where the $\mathbf{x}_i$ are the possible values of $\mathbf{X}$.

3. Use the fact that the Lebesgue-Stieltjes integral has the additivity property to show that for any jointly distributed X and Y we have

$$E(X+Y) = E(X) + E(Y)$$

Hint: Use Section 4.2, Corollary 2, again.

4. Let X and Y both defined on $(S, \mathcal{S}, P)$ be independent and have Riemann integrable density functions. Show that

$$E_S(XY) = E_S(X)E_S(Y)$$

Hint: In Section 4.2, Corollary 2, let $g(x, y) = xy$; use the fact that $f_{X, Y}(x, y) = f_X(x) f_Y(y)$ for X and Y independent.

5. Let X and Y be as in the preceding problem, and further assume that $E_S(X)= E_S(Y) = 0$. Show that

$$E_S[(X + Y)^2] = E_S(X^2) + E_S(Y^2)$$

Is $E_S[(X + Y)^n] = E_S(X^n) + E_S(Y^n)$ true for $n = 3$? For $n = 4$?

4.3 NUMERICAL CHARACTERIZATIONS OF A RANDOM VARIABLE

The expected value $E(X)$ of a random variable X provides an average value for the distribution P_X. If we visualize P_X in terms of (probability) mass, then we ought to visualize $E(X)$ as the center of mass distributed according to P_X. It is important to note that the definition of $E(X)$ depends on the distribution of X, and not so much on X itself. In other words, $E(X)$ is not so much a numerical characterization of X as of P_X. However, whatever characterizes P_X also inevitably characterizes X, therefore $E(X)$ certainly may also be regarded as a numerical characterization of X.

4.3.1 DEFINITION

We define the *mean value* of the distribution P_X, and hence of X, by

$$\mu_X = E(X)$$

In particular, if X has a discrete distribution, then

$$\mu_X = \sum_i x_i p_i$$

and if X has a density function, then

$$\mu_X = \int_{-\infty}^{\infty} x f_X(x)\, dx$$

Example 1 Find the mean value of an (n, p) binomial distribution.

Solution If X counts the number of heads in n tosses of a p-coin, then X has an (n, p) binomial distribution. Now according to Section 4.2, Example 1, or Problem 4.1.14, $E(X) = np$, consequently the mean value of (n, p) binomial distribution is np.

Example 2 Find the mean value of the standard normal distribution.

Solution Let X have such a distribution,

$$f_X(x) = \frac{1}{\sqrt{2\pi}} e^{-(1/2)x^2}$$

Then

$$\mu_X = E(X) = \int_{-\infty}^{\infty} \frac{1}{\sqrt{2\pi}}\, xe^{-(1/2)x^2}\, dx$$

This integral is 0 since the integrand is an odd function, hence $\mu_X = 0$.

The basic properties of mathematical expectation such as

(a) $$E(aX + bY) = aE(X) + bE(Y)$$

(b) $$E(X \cdot Y) = E(X)E(Y) \qquad \text{for independent } X \text{ and } Y$$

can be translated into the corresponding properties for μ. Thus we have

(i) $\mu_{aX+bY} = a\mu_X + b\mu_Y$

(ii) $\mu_{X \cdot Y} = \mu_X \mu_Y \qquad$ for independent X and Y

As a special case of (i) in which $Y = 1$ (a constant random variable) we have

$$\mu_{aX+b} = a\mu_X + b$$

This property may come in handy for situations like that in the following example.

Example 3 Calculate the mean value of Y having a (μ, σ) normal distribution.

Solution We have seen that if Y is (μ, σ) normal then

$$X = \frac{Y - \mu}{\sigma}$$

is standard normal (Section 3.3, Example 3), and by the Example 2 of this section $\mu_X = 0$. Now

$$Y = \sigma X + \mu$$

hence

$$\mu_Y = \mu_{\sigma X + \mu} = \sigma \mu_X + \mu = \sigma \cdot 0 + \mu = \mu$$

This explains the choice of the letter μ for (μ, σ) normal distribution.

The fact that $\mu_{aX} = a\mu_X$ or that $\mu_{X+b} = \mu_X + b$ is not surprising at all if we recall the interpretation of μ_X as the center of mass.

Although μ_X gives some sort of indication regarding the distribution of X, it hardly distinguishes between different distributions. For example, if X is evenly distributed between the two points ± 1, then clearly $\mu_X = 0$; likewise if Y is evenly distributed between the two points ± 10, then again $\mu_Y = 0$; yet Y shows a marked degree of "scatter" or "spread" as compared to X. In order to numerically characterize the degree of spread of probability mass about its mean value, we shall consider the probability-weighted average of deviation of values of random variable from its mean value. We introduce the following definition.

4.3.2 DEFINITION

We define the *variance* of the distribution P_X, and hence of X, by

$$\text{var}(X) = E[(X - \mu_X)^2]$$

In other words, variance of X is the probability-weighted average of the squared deviation of X from its mean value. The squaring of $X - \mu_X$ is necessary to prevent cancellations between positive and negative deviations; however, it does have an inflating effect. For example, for X and Y above, which had two-point distributions, we would have

$$\text{var}(X) = (1 - 0)^2 \tfrac{1}{2} + (-1 - 0)^2 \tfrac{1}{2} = 1$$
$$\text{var}(Y) = (10 - 0)^2 \tfrac{1}{2} + (-10 - 0)^2 \tfrac{1}{2} = 100$$

In order to nullify this inflating effect of squaring we consider the following definition.

4.3.3 DEFINITION

We define the *standard deviation* of the distribution P_X, hence of X, by

$$\sigma_X = \sqrt{\text{var}(X)}$$

Thus, X and Y above would have more proportionate standard deviations: $\sigma_X = 1$ and $\sigma_Y = 10$. Unfortunately, standard deviation is less amenable to mathematical manipulations than variance; thus most theorems are formulated in terms of variance.

Example 4 A coin is tossed once. Let X denote the number of heads and calculate $\text{var}(X)$ and σ_X.

Solution Assuming the coin has the probability p of falling heads, we have

$$\mu_X = 1 \cdot p + 0 \cdot q = p$$

hence

$$\begin{aligned}
\text{var}(X) &= E[(X - \mu_X)^2] \\
&= (1 - p)^2 \cdot p + (0 - p)^2 \cdot q \\
&= q^2 \cdot p + p^2 \cdot q \\
&= pq(p + q) = pq
\end{aligned}$$

and

$$\sigma_X = \sqrt{pq}$$

Example 5 Let X have a uniform distribution over the interval $[-1, 1]$. Calculate $\text{var}(X)$ and σ_X.

Solution By symmetry of distribution obviously

$$\mu_X = 0$$

Now

$$\text{var}(X) = E[(X - \mu_X)^2]$$

$$= \int_{-\infty}^{\infty} (x - 0)^2 f_X(x)\, dx$$

but $f_X(x) = \frac{1}{2}$ for $-1 \leq x \leq 1$, hence

$$\text{var}(X) = \frac{1}{2} \int_{-1}^{1} x^2\, dx = \frac{1}{2} \frac{x^3}{3} \Big|_{-1}^{1} = \frac{1}{6}(1 + 1) = \frac{1}{3}$$

Variance is essentially a "derived" operator since $\text{var}(X) = E[(X - \mu_X)^2]$ and E is an operator. The basic properties of variance can thus be derived from those of mathematical expectation E. These are

(i) $\text{var}(X + b) = \text{var}(X)$ for any   constant b
(ii) $\text{var}(aX) = a^2 \text{var}(X)$ so that $\sigma_{aX} = |a|\sigma_X$
(iii) $\text{var}(X + Y) = \text{var}(X) + \text{var}(Y)$ for independent X and Y

Proofs for (i) and (ii) can be done as simple exercises. Let us examine (iii).

$$\text{var}(X + Y) = E[(X + Y - \mu_{X+Y})^2]$$
$$= E\{[(X - \mu_X) + (Y - \mu_Y)]^2\}$$
$$= E[(X - \mu_X)^2 + (Y - \mu_Y)^2 + 2(X - \mu_X)(Y - \mu_Y)]$$
$$= E[(X - \mu_X)^2] + E[(Y - \mu_Y)^2] + 2E[(X - \mu_X)(Y - \mu_Y)]$$
$$= \text{var}(X) + \text{var}(Y) + 2E[(X - \mu_X)(Y - \mu_Y)]$$

So far we have not assumed that X and Y are independent. Now if X and Y are independent, so are $X - \mu_X$ and $Y - \mu_Y$, consequently

$$E[(X - \mu_X)(Y - \mu_Y)] = E(X - \mu_X)E(Y - \mu_Y) = 0 \cdot 0$$

and (iii) is established. It is now a simple matter of mathematical induction to arrive at

$$\text{var}(X_1 + X_2 + \cdots + X_n) = \text{var}(X_1) + \text{var}(X_2) + \cdots + \text{var}(X_n)$$

for an independent class of random variables.

Example 6 Find the variance of an (n, p) binomial distribution.

Solution If X denotes the number of heads in n tosses of a p-coin, X has an (n, p) binomial distribution. To calculate $\text{var}(X)$, we express X as

$$X = X_1 + X_2 + \cdots + X_n$$

where X_i denotes the number of heads for the ith toss only. Since X_i are mutually independent, by property (iii) we obtain

$$\text{var}(X) = \text{var}(X_1) + \text{var}(X_2) + \cdots + \text{var}(X_n)$$

but by Example 4, $\text{var}(X_i) = pq$, consequently

$$\text{var}(X) = npq$$

Example 7 Find the variance of the standard normal distribution and the (μ, σ) normal distribution.

Solution Let X have the normal density

$$f_X(x) = \frac{1}{\sqrt{2\pi}}\, e^{-(1/2)x^2}$$

then since $\mu_X = 0$ by Example 2 of this section

$$\text{var}(X) = E[(X - \mu_X)^2]$$
$$= E(X^2)$$
$$= \int_{-\infty}^{\infty} x^2 \frac{1}{\sqrt{2\pi}}\, e^{-(1/2)x^2}\, dx$$

To evaluate the integral we integrate by parts. Let

$$u = x \qquad\qquad \text{and so}\quad du = dx$$
$$dv = xe^{-(1/2)x^2}\, dx \qquad \text{and so}\quad v = -e^{-(1/2)x^2}$$

then

$$\int_{-\infty}^{\infty} x^2 e^{-(1/2)x^2}\, dx = -xe^{-(1/2)x^2}\Big|_{-\infty}^{\infty} + \int_{-\infty}^{\infty} e^{-(1/2)x^2}\, dx$$

$$= 0 + \sqrt{2\pi}$$

Consequently

$$\text{var}(X) = 1$$

Next, let Y have a (μ, σ) normal density, then

$$X = \frac{Y - \mu}{\sigma}$$

has a standard normal density. Consequently

$$\begin{aligned} \text{var}(Y) &= \text{var}(\sigma X + \mu) \\ &= \sigma^2 \, \text{var}(X) \\ &= \sigma^2 \end{aligned}$$

so that $\sigma_Y = \sigma$, and this explains the choice of the letter σ in the normal density function.

Given a random variable X, we can consider other numerical characterizations beside μ_X, σ_X, and σ_X^2 (σ^2 being an alternative notation for var). Also, since σ_X and σ_X^2 are examples of conceptually dependent numerical characterizations, this suggests that we look only for conceptually independent numerical characterizations.

4.3.4 DEFINITION

We define the *kth moment* of the distribution P_X, and hence of X, by

$$m_X^{(k)} = E(X^k) \qquad \text{for} \quad k = 1, 2, 3, \ldots$$

These moments are fundamental in that μ_X and σ_X^2 can be derived from them. First of all, quite trivially

$$\mu_X = m_X^{(1)}$$

As for σ_X^2, we have

$$\begin{aligned} \sigma_X^2 &= E[(X - \mu_X)^2] \\ &= E[X^2 - 2\mu_X \cdot X + \mu_X^2] \\ &= E(X^2) - 2\mu_X \cdot E(X) + \mu_X^2 \\ &= m_X^{(2)} - \mu_X^2 \end{aligned}$$

In other words, the variance is equal to the second moment minus the square of the first moment. This computation formula for variance may prove quite useful on occasions.

Example 8 Find the mean value and the variance of the μ-Poisson distribution.

Solution Let X have a μ-Poisson distribution

$$P_X(k) = \frac{e^{-\mu}\mu^k}{k!}$$

then

$$\begin{aligned} \mu_X &= \sum_{k=0}^{\infty} \frac{k e^{-\mu}\mu^k}{k!} \\ &= \mu \sum_{k=1}^{\infty} \frac{e^{-\mu}\mu^{k-1}}{(k-1)!} \end{aligned}$$

Replacing $k - 1$ by k throughout, we obtain

$$\mu_X = \mu \sum_{k=0}^{\infty} \frac{e^{-\mu}\mu^k}{k!} = \mu$$

To evaluate $\sigma_X^2 = m_X^{(2)} - \mu^2$, we consider first

$$m_X^{(2)} = E(X^2) = \sum_{k=0}^{\infty} \frac{k^2 e^{-\mu}\mu^k}{k!}$$

Noting $k^2 = k(k - 1) + k$, we write

$$m_X^{(2)} = \sum_{k=2}^{\infty} k(k - 1) \frac{e^{-\mu}\mu^k}{k!} + \sum_{k=0}^{\infty} k \frac{e^{-\mu}\mu^k}{k!}$$

$$= \mu^2 \sum_{k=2}^{\infty} \frac{e^{-\mu}\mu^{k-2}}{(k - 2)!} + \mu$$

Again, replacing $k - 2$ by k throughout, we obtain

$$m_X^{(2)} = \mu^2 + \mu$$

Consequently,

$$\sigma_X^2 = \mu$$

It is conceivable that a random variable may not have all its moments defined (see Problem 4.3.6). However, it can be shown (see Problem 4.3.7) that if X has a kth moment it has all the lower moments. In Chapter 6 we shall introduce the moment-generating function M_X, which "embodies" all the moments of X whenever they exist.

4.3 PROBLEMS

1. Calculate the mean and the variance of the exponential distribution:

$$f_X(x) = \lambda e^{-\lambda x} \qquad \text{for} \quad x > 0$$

2. Calculate the mean and the variance of the triangular distribution over the interval $[a, b]$.

3. Calculate the mean and the variance of the geometric distribution:

$$P_X(k) = pq^{k-1} \qquad \text{for} \quad k = 1, 2, 3, \ldots$$

4. If X is uniformly distributed over the interval $[0, 1]$, show that

$$m_X^{(k)} = \frac{1}{k + 1} \qquad \text{for} \quad k = 1, 2, 3, \ldots$$

From this, show that

$$\sigma_X^2 = \tfrac{1}{12}$$

5. If X is uniformly distributed over the interval $[-1, 1]$, show that

$$m_X^{(k)} = \begin{cases} 0 & \text{for odd } k \\ \dfrac{1}{k+1} & \text{for even } k \end{cases}$$

Use this result to show

$$\text{var}(X^n) = \begin{cases} \dfrac{1}{2n+1} & \text{for odd } n \\ \dfrac{1}{2n+1} - \dfrac{1}{(n+1)^2} & \text{for even } n \end{cases}$$

6. Show that the second moment does not exist for X having a Cauchy distribution

$$f_X(x) = \frac{1}{\pi} \frac{a}{a^2 + x^2} \qquad \text{for} \quad -\infty < x < \infty$$

Hint: Integrate with change of variable $x = a \tan \theta$.

7. Given a random variable X prove that

$$E(|X|^{m-1}) \leqq E(|x|^m) + 1$$

by first showing that for any real number x we have

$$|x|^{m-1} \leqq |x|^m + 1$$

Hint: Consider two cases, $|x| \geqq 1$ and $|x| \leqq 1$.

8. Show that $\text{var}(X) = 0$ if and only if X has a one-point distribution, that is, X assumes the same constant value almost certainly, $X = c$ (with probability 1).

9. If X has a (λ, r) gamma distribution (cf. Definition 3.2.2), X is the sum of r independent random variables all having λ exponential distributions. Find the variance of a λ exponential distribution; from this, find the variance of a (λ, r) gamma distribution.

10. A coin with probability p of falling heads is tossed repeatedly until a total of r heads appear. Calculate the variance of the random variable X denoting the total number of tosses. Hint: Express X as the sum of r independent random variables all having identical geometric distributions.

4.4 NUMERICAL CHARACTERIZATIONS OF A RANDOM VECTOR

We recall from the preceding section that various numerical characterizations for a random variable X were obtained by (1) considering $E[g(X)]$ for various Borel functions g and (2) algebraically combining these numerical values obtained in (1). For instance, the first and the second moments of X were obtained from $E(X)$ and $E(X^2)$ by letting $g(x) = x$ and $g(x) = x^2$ respectively, then the variance of X was obtained from the combination $E(X^2) - E(X)^2$. Now there is no reason why we cannot do the same for random vectors to obtain numerical characterizations that may prove useful. Given an n-dimensional random vector $\mathbf{X}$ let us consider $E[g(\mathbf{X})]$ for various Borel functions g and some algebraic combinations based on these numerical values.

4.4.1 DEFINITION

Given a random vector $\mathbf{X} = (X_1, X_2, \ldots, X_n)$, its *jth marginal mean value* is defined by

$$\mu_j(\mathbf{X}) = E(X_j)$$

In other words we have taken $g(x_1, x_2, \ldots, x_n) = x_j$ to arrive at this particular numerical characterization of X, which of course is nothing but the mean value of the random variable X_j.

4.4.2 DEFINITION

Given a random vector $\mathbf{X} = (X_1, X_2, \ldots, X_n)$ its $(j, k)th$ *covariance* is defined by

$$\sigma_{jk}(\mathbf{X}) = E[(X_j - \mu_j)(X_k - \mu_k)]$$

where μ_j and μ_k are abbreviations for $\mu_j(\mathbf{X})$ and $\mu_k(\mathbf{X})$ respectively.

Note that since $(X_j - \mu_j)(X_k - \mu_k) = X_j X_k - \mu_j X_k - \mu_k X_j + \mu_j \mu_k$ we have the formula

$$\sigma_{jk}(\mathbf{X}) = E(X_j X_k) - E(X_j)E(X_k)$$

Note also that (j, j) covariance of $\mathbf{X}$ is none other than the variance of X_j since

$$\sigma_{jj}(\mathbf{X}) = E(X_j X_j) - E(X_j)E(X_j)$$
$$= E(X_j^2) - E(X_j)^2 = \sigma^2(X_j)$$

Since the meaning of $\mu_j(\mathbf{X})$ is already clear, let us consider the meaning of $\sigma_{jk}(\mathbf{X})$. For example, what interpretation can we make in regard to X_j and X_k if $\sigma_{jk}(\mathbf{X})$ is positive, negative, or 0? Perhaps the following example will reveal the meaning of σ_{jk}. Incidentally, whenever convenient we shall adopt the more customary notation $\mathrm{cov}(X_j, X_k)$ for $\sigma_{jk}(\mathbf{X})$.

Example 1 Let the random vector (X, Y) be distributed in three different ways as follows (see Fig. 4.4.1 where black dots represent probability mass). Calculate $\mathrm{cov}(X, Y)$ in each case.

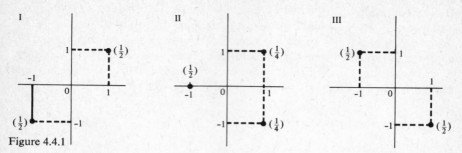

Figure 4.4.1

Solution Using the formula $\text{cov}(X, Y) = E(XY) - E(X)E(Y)$ we calculate

(I) $\text{cov}(X, Y) = [(1)(1)(\frac{1}{2}) + (-1)(-1)(\frac{1}{2})] - 0 \cdot 0 = 1$
(II) $\text{cov}(X, Y) = [(1)(1)(\frac{1}{4}) + (1)(-1)(\frac{1}{4}) + (-1)(0)(\frac{1}{2})] - 0 \cdot 0 = 0$
(III) $\text{cov}(X, Y) = [(1)(-1)(\frac{1}{2}) + (-1)(1)(\frac{1}{2})] - 0 \cdot 0 = -1$

As illustrated above, $\text{cov}(X, Y) = E[(X - \mu_X)(Y - \mu_Y)]$ easily ends up positive if, whenever X assumes values larger than μ_X, Y also assumes values larger than μ_Y and, whenever X assumes values smaller than μ_X, Y also assumes values smaller than μ_Y (Case I). On the other hand, $\text{cov}(X, Y)$ easily takes a negative value if, whenever X takes values larger than μ_X, Y takes values smaller than μ_Y and, whenever X takes values smaller than μ_X, Y takes values larger than μ_Y (Case III). In between these two clear-cut situations, cancellations can take place among positive and negative values of $(X - \mu_X)(Y - \mu_Y)$ so that its probability-weighted average $E[(X - \mu_X)(Y - \mu_Y)]$ may be barely positive, or barely negative, or vanish altogether (Case II).

4.4.3 DEFINITION

If two random variables X and Y are defined on the same probability space so that

$$\text{cov}(X, Y) = 0$$

we say that X and Y are *uncorrelated*. By the same token if a random vector $(X_1, X_2, \ldots, X_n)$ is such that

$$\sigma_{jk} = \text{cov}(X_j, X_k) = 0$$

for all $j \neq k$, we say that the random vector has (*pairwise*) *uncorrelated* components.

Clearly if X and Y are uncorrelated this can be taken to mean that X and Y fluctuate about their respective mean values quite independently; however, this does not necessarily mean that X and Y are stochastically independent, as is clear from Case II of Example 1 of this section. On the other hand if X and Y are stochastically independent, they are certainly uncorrelated, as asserted in the following proposition.

4.4.1 PROPOSITION

If X and Y are independent, then X and Y are uncorrelated.

PROOF

We need only evaluate $\text{cov}(X, Y) = E(XY) - E(X)E(Y)$. But since X and Y are independent, $E(XY) = E(X)E(Y)$. It follows that $\text{cov}(X, Y) = 0$.

In short, stochastic independence is a much more stringent condition than uncorrelatedness. In the preceding section we have established that variances of random variables are additive if these random variables are pairwise independent. A quick review of the proof shows however that we did not really need independence since uncorrelatedness will suffice. In other words we have the following proposition.

4.4.2 PROPOSITION

If the random vector $(X_1, X_2, \ldots, X_n)$ has uncorrelated components, then

$$\text{var}\left(\sum_j X_j\right) = \sum_j \text{var}(X_j)$$

PROOF

We proceed somewhat mechanically.

$$\text{var}\left(\sum_j X_j\right) = E\left[\left(\sum_j X_j\right)^2\right] - \left[E\left(\sum_j X_j\right)\right]^2$$

$$= \left[\sum_j E(X_j^2) + \sum_{j<k} 2E(X_j X_k)\right] - \sum_j [E(X_j)]^2 - \sum_{j<k} 2E(X_j)E(X_k)$$

$$= \sum_j \text{var}(X_j) + 2\sum_{j<k} \text{cov}(X_j, X_k)$$

$$= \sum_j \text{var}(X_j)$$

Let us resume our discussion of covariance: $\text{cov}(X, Y)$ is a numerical value assigned to a pair of random variables X and Y, that is, cov operates on a pair (X, Y) to produce a number $\text{cov}(X, Y)$. In other words, cov is a binary operator. As a binary operator cov has the following convenient properties:

(i) $\text{cov}(X, Y) = \text{cov}(Y, X)$
(ii) $\text{cov}(aX, Y) = a\,\text{cov}(X, Y)$
(iii) $\text{cov}(X + X', Y) = \text{cov}(X, Y) + \text{cov}(X', Y)$

These basic properties of cov can be easily deduced from the basic properties of mathematical expectation E (see Problem 4.4.9).

Covariance behaves well mathematically just as variance does, but again is somewhat "inflated" in a sense just as variance is. Out of covariance we will now extract *correlation coefficient* just as we extracted standard deviation out of variance. For convention's sake, let us agree to say that X and Y tend to "resonate" if $\text{cov}(X, Y) > 0$ and that X and Y tend to "interfere" if $\text{cov}(X, Y) < 0$. Now given two random vectors (X, Y) and (X', Y'), if $\text{cov}(X, Y) > \text{cov}(X', Y') > 0$, should we interpret this as meaning that X and Y have a stronger tendency to resonate than X' and Y'? We would hesitate to do so since it is possible that $\text{cov}(X, Y)$ is greater than $\text{cov}(X', Y')$ for the trivial reason that X and Y as a whole assume larger values than X' and Y'. For example, X and Y may represent the height and the weight of an American male measured in inches and ounces while X' and Y' represent the same measured in feet and pounds; or consider the following example, in which $\text{cov}(X, Y) > \text{cov}(X', Y')$ and yet we can hardly regard X and Y as having a stronger tendency to resonate than X' and Y'.

Example 2 Calculate $\text{cov}(X, Y)$ and $\text{cov}(X', Y')$ if (X, Y) and (X', Y') are distributed respectively as in Fig. 4.4.2.

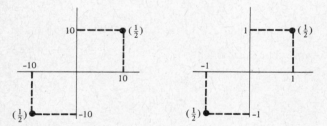

Figure 4.4.2

Solution Straightforward calculations give

$$\text{cov}(X, Y) = (10)(10)(\tfrac{1}{2}) + (-10)(-10)(\tfrac{1}{2}) - 0 \cdot 0 = 100$$
$$\text{cov}(X', Y') = (1)(1)(\tfrac{1}{2}) + (-1)(-1)(\tfrac{1}{2}) - 0 \cdot 0 = 1$$

It should be very clear by now that $\text{cov}(X, Y)$ as a gauge of resonance (or correlation) between X and Y contains extraneous factors which must be eliminated to give way to a "truer" gauge of correlation between X and Y. Accordingly we consider the following definition.

4.4.4 DEFINITION

Given a random vector $\mathbf{X} = (X_1, X_2, \ldots, X_n)$, its $(i, j)th$ *correlation coefficient* is defined by

$$\rho_{ij}(\mathbf{X}) = E\left[\left(\frac{X_i - \mu_i}{\sigma_i}\right)\left(\frac{X_j - \mu_j}{\sigma_j}\right)\right] = \frac{\sigma_{ij}(\mathbf{X})}{\sigma_i \sigma_j}$$

where σ_i and σ_j are the standard deviations of X_i and X_j, respectively. (Note ρ_{ij} is not defined unless $\sigma_i \sigma_j > 0$.) Alternatively we may also write

$$\rho(X_i, X_j) = \frac{\text{cov}(X_i, X_j)}{\sigma_i \sigma_j}$$

or

$$\text{cov}(X_i, X_j) = \sigma_i \sigma_j \rho(X_i, X_j)$$

or briefly

$$\sigma_{ij} = \sigma_i \sigma_j \rho_{ij}$$

Example 3 Calculate the correlation coefficient $\rho(X, Y)$ if (X, Y) is distributed evenly among the four points indicated in Fig. 4.4.3.

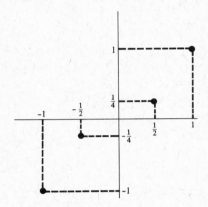

Figure 4.4.3

Solution Since obviously $\mu_X = \mu_Y = 0$, we have

$$\text{cov}(X, Y) = 1 \cdot 1 \cdot \tfrac{1}{4} + \tfrac{1}{2} \cdot \tfrac{1}{4} \cdot \tfrac{1}{4} + (-1)(-1)\tfrac{1}{4} + (-\tfrac{1}{2})(-\tfrac{1}{4})\tfrac{1}{4} = \tfrac{9}{16}$$

and

$$\sigma_X^2 = 1^2 \cdot \tfrac{1}{4} + (\tfrac{1}{2})^2 \cdot \tfrac{1}{4} + (-1)^2\tfrac{1}{4} + (-\tfrac{1}{2})^2\tfrac{1}{4} = \tfrac{5}{8}$$

$$\sigma_Y^2 = 1^2 \cdot \tfrac{1}{4} + (\tfrac{1}{4})^2 \cdot \tfrac{1}{4} + (-1)^2\tfrac{1}{4} + (-\tfrac{1}{4})^2\tfrac{1}{4} = \tfrac{17}{32}$$

Consequently,

$$(X, Y) = \tfrac{9}{16} \sqrt{\tfrac{5}{8}} \sqrt{\tfrac{17}{32}} \doteq 0.98$$

Accepting the correlation coefficient as a gauge of the extent of correlation between two jointly distributed random variables, we now investigate just how large (or how small) a correlation coefficient can be and what it means for the correlation coefficient to assume the maximum (or minimum) value. According to Definition 4.4.4 a correlation coefficient is nothing but the expected value of the product of two standardized random variables, thus in order to find the possible extreme values for correlation coefficients we need only examine the totality $E(XY)$ where X and Y range through the standardized random variables. Now, for standardized X and Y we have (cf. page 162 and Definition 4.4.4)

$$\operatorname{var}(X + Y) = \operatorname{var}(X) + \operatorname{var}(Y) + 2\sigma_X\sigma_Y\rho(X, Y)$$
$$= 1 + 1 + 2\rho(X, Y)$$

But since variance is always nonnegative, $\operatorname{var}(X + Y) \geqq 0$, therefore we have

$$2 + 2\rho \geqq 0$$

Consequently,

$$\rho \geqq -1$$

Thus ρ cannot be too small, being at least -1. Now this means that ρ cannot be too large either, for if $\rho(X, Y)$ were greater than 1 for some X and Y, then $\rho(-X, Y)$ would have to be less than -1 since $\rho(-X, Y) = -\rho(X, Y)$. We have thus established the following proposition.

4.4.3 PROPOSITION
Given a random vector $\mathbf{X} = (X_1, X_2, \ldots, X_n)$ we have for all $i, j = 1, 2, \ldots, n$

$$|\rho_{ij}(\mathbf{X})| \leqq 1$$

The question, what it means for $\rho_{ij}(\mathbf{X})$ to be exactly 1 or -1, is answered by the following proposition.

4.4.4 PROPOSITION
Given two jointly distributed random variables X and Y, we have
(i) $\rho(X, Y) = 1$ if and only if $Y = aX + b$ with $a > 0$.
(ii) $\rho(X, Y) = -1$ if and only if $Y = aX + b$ with $a < 0$.

PROOF
First we prove (i) in detail. Let X' and Y' be respectively the standardizations of X and Y, then $\rho(X, Y) = \rho(X', Y')$. Therefore if $\rho(X, Y) = 1$, we have

$$\operatorname{var}(X' - Y') = \operatorname{var}(X') + \operatorname{var}(Y') - 2\rho(X', Y')$$
$$= 1 + 1 - 2 = 0$$

Now a random variable cannot have zero variance unless that random variable has a one-point distribution, that is, is equal to a constant (with probability 1). Hence $X' - Y' = c$ (with probability 1). Substituting $X' = (X - \mu_1)/\sigma_1$ and $Y' = (Y - \mu_2)/\sigma_2$ above, we obtain after simplification

$$Y = aX + b \qquad \text{(with probability 1)}$$

with $a = \sigma_2/\sigma_1$ and b an expression involving μ_1, μ_2, σ_1, σ_2, and c. Thus the joint distribution of X and Y is such that the entire probability mass is distributed on the line $y = ax + b$.

Conversely now, suppose $Y = aX + b$, $a > 0$ (with probability 1), then simple substitutions show $Y' = a'X' + b'$ where $a' > 0$ and b' are constants involving $a, \sigma_1, \sigma_2, \mu_1, \mu_2$. Now, a' has to be 1 since both X' and Y' being standardizations of X and Y have variances 1. Hence $Y' = X' + b'$, and $\mathrm{var}(X' - Y') = \mathrm{var}(-b') = 0$ follows. On the other hand

$$\mathrm{var}(X' - Y') = \mathrm{var}(X') + \mathrm{var}(Y') - 2\,\mathrm{cov}(X', Y')$$
$$= 1 + 1 - 2\rho(X', Y')$$

Consequently, $\rho(X', Y') = 1$, hence $\rho(X, Y) = 1$ also, completing the proof of (i).

As for (ii), we need only observe that $\rho(X, Y) = -1$ is equivalent to $\rho(-X, Y) = 1$, which in turn is equivalent to $Y = a(-X) + b = (-a)X + b$ according to (i), where $a > 0$ and so $-a < 0$.

The preceding proposition shows that $|\rho_{ij}(\mathbf{X})| = 1$ signifies linear dependence of X_i and X_j. Note that $X_j = aX_i + b$ is equivalent to $X_i = (1/a)X_j - (b/a) = a'X_j + b'$. If X_i and X_j are functionally dependent but not linearly so, then of course we have no reason to expect $\rho_{ij}(\mathbf{X}) = 1$. For instance, consider the following example.

Example 4 Let the jointly distributed random variables X and Y be such that $Y = X^n$, and the marginal distribution of X is uniform over the interval $[-1, 1]$. Calculate $\rho(X, Y)$.

Solution First we calculate $\mathrm{cov}(X, Y)$.

$$\mathrm{cov}(X, X^n) = E(X \cdot X^n) - E(X)E(X^n) = E(X^{n+1}) - 0 \cdot E(X^n)$$

$$= \int_{-1}^{1} x^{n+1}\, dx = \left.\frac{x^{n+2}}{n+2}\right|_{-1}^{1} = \begin{cases} \dfrac{2}{n+2} & \text{if } n \text{ is odd} \\[2mm] 0 & \text{if } n \text{ is even} \end{cases}$$

Thus, for even n, $\mathrm{cov}(X, X^n) = 0$, so that X and X^n are uncorrelated though indeed they are highly dependent. As for odd n, we must divide $\mathrm{cov}(X, Y)$ by $\sigma_X \cdot \sigma_Y$ to obtain $\rho(X, Y)$. Now

$$\sigma_X^2 = \int_{-1}^{1} x^2\, dx = \left.\frac{x^3}{3}\right|_{-1}^{1} = \frac{2}{3}$$

$$\sigma_Y^2 = \int_{-1}^{1} x^{2n}\, dx = \left.\frac{x^{2n+1}}{2n+1}\right|_{-1}^{1} = \frac{2}{2n+1}$$

Consequently we have

$$(X, X^n) = \frac{\dfrac{2}{n+2}}{\sqrt{\dfrac{2}{3}}\sqrt{\dfrac{2}{2n+1}}} = \frac{\sqrt{3(2n+1)}}{n+2}$$

Note that if $n = 1$, $\rho(X, X^n) = 1$; and as n approaches ∞, $\rho(X, X^n)$ approaches 0, thus for large values of n, X and X^n become nearly uncorrelated.

From this example and the preceding proposition we see that $|\rho(X, Y)|$ is a measure of linear dependence between X and Y; the closer $|\rho(X, Y)|$ is to 1, the more X and Y approach linear dependence. In Example 3 of this section the correlation would be even closer to 1 if the four mass points were more nearly on a straight line. Finally, we point out that if X and Y are binormally distributed according to the density function

$$f_{X, Y}(x, y) = \frac{1}{2\pi} \exp\left[\frac{1}{2(1 - \rho^2)} (x^2 - 2\rho xy + y^2)\right]$$

then $\rho(X, Y)$ turns out to be exactly ρ appearing in the density function, and this of course explains the choice of the letter ρ. We shall not burden ourselves with the tedious calculation

$$\int_{-\infty}^{\infty} \int_{-\infty}^{\infty} xy f_{X, Y}(x, y) \, dy \, dx = \rho$$

4.4 PROBLEMS

1. Given a random vector $\mathbf{X} = (X_1, X_2, \ldots, X_n)$,
    a. If μ_i denotes the ith marginal mean value $\mu_i(\mathbf{X}) = \mu(X_i)$ what physical interpretation can you give to the *mean vector* $\mu(\mathbf{X}) = (\mu_1, \mu_2, \ldots, \mu_n)$?
    b. If σ_{ij} denotes the (i, j) covariance $\sigma_{ij}(\mathbf{X}) = \operatorname{cov}(X_i, X_j)$ what mathematical properties do you observe in the *covariance matrix* $\sigma(\mathbf{X}) = (\sigma_{ij})$?

2. Given jointly distributed X and Y, $\rho(X, Y)$ is not defined unless $\operatorname{var}(X) > 0$ and $\operatorname{var}(Y) > 0$. Assuming that both $\operatorname{cov}(X, Y)$ and $\rho(X, Y)$ are defined, show that $\operatorname{cov}(X, Y) = 0$ if and only if $\rho(X, Y) = 0$.

3. Show that if either X or Y is a constant, then $\operatorname{cov}(X, Y) = 0$ and $\rho(X, Y)$ is not defined.

4. Show that if the joint density function of X and Y satisfies the condition

$$f(x, y) = f(-x, y) \quad [\text{or } f(x, y) = f(x, -y)]$$

then $\rho(X, Y) = \operatorname{cov}(X, Y) = 0$.

5. Calculate $\rho(X, X^{1/3})$ if X is uniformly distributed over the interval $[-1, 1]$.

6. Show that if (X, Y) has a two-point distribution and $\operatorname{cov}(X, Y) = 0$, then X and Y must be independent. Hint: Assume without loss of generality that $E(X) = E(Y) = 0$. From $E(XY) = p_1 x_1 y_1 + p_2 x_2 y_2 = 0$ deduce, for example, $y_1/y_2 = -p_2 x_2/p_1 x_1$; then note $p_2 x_2/p_1 x_1 = 1$, since $E(X) = 0$; hence $y_1 = y_2$; and so on.

7. Show that for any jointly distributed X and Y we have

$$\rho(aX + b, cY + d) = \rho(X, Y)$$

8. Optional. Show $\text{cov}(X, \ Y) = \text{cov}(Y, \ X)$, $\text{cov}(aX, \ Y) = a \ \text{cov}(X, \ Y)$, and $\text{cov}(X' + X'', \ Y) = \text{cov}(X', \ Y) + \text{cov}(X'', \ Y)$, then combine these to show the *bilinearity* of cov:

$$\text{cov}\left(\sum_i a_i X_i, \sum_j b_j Y_j\right) = \sum_{i,j} a_i b_j \text{cov}(X_i, \ Y_j)$$

9. Optional. Use the result of Problem 4.4.8 to show

$$\text{var}\left(\sum_i a_i X_i\right) = \sum_i a_i^2 \text{var}(X_i) + 2 \sum_{i<j} a_i a_j \text{cov}(X_i, \ X_j)$$

and hence in particular

$$\text{var}\left(\sum_i a_i X_i\right) = \sum_i a_i^2 \text{var}(X_i)$$

if the X_i are pairwise uncorrelated.

10. Schwarz inequality. Given jointly distributed X and Y, assuming $E(X^2) < \infty$ and $E(Y^2) < \infty$, show that

$$E(XY) \leq \sqrt{E(X^2)} \sqrt{E(Y^2)}$$

Hint: Use Proposition 4.4.3.

Chapter 5
CONDITIONAL
DISTRIBUTIONS AND
EXPECTATIONS

In this chapter, we continue the analysis of random vectors begun in the latter half of Chapter 3 and the last section of Chapter 4. Essentially, Sections 5.1 and 5.2 are continuations of Chapter 3, and Sections 5.3 and 5.4 are continuations of Chapter 4.

In Section 5.1 (Conditional Distributions), the idea of conditional distribution of a random variable given another is presented somewhat intuitively in terms of probability mass. Unfortunately, the formal definition of conditional distribution may in some cases create difficulty. This difficulty is caused essentially by our insistence that conditional distribution be defined within the context of the joint distribution. The solution therefore is to consider what we call *preconditional distributions*, which are modeled directly from the random experiment. Certain simple regularity conditions are then imposed on these preconditional distributions to ensure that they "generate" a joint distribution. Such an approach is clearly more in line with the thinking that conditional probabilities are useful only in as much as they lead to determinations of (nonconditional) probabilities.

In Section 5.2 (Marginal and Conditional Density Functions), we consider the same questions as in Section 5.1, except that we now confine ourselves to absolutely continuous random vectors. Here everything happens in much neater forms, and we are able to obtain a number of useful formulas involving joint density, marginal density, and conditional density functions.

In Section 5.3 (Conditional Expected Values), we consider numerical characteristics of a random vector, such as conditional mean, conditional variance, and regression coefficients, that are of particular importance in statistics. These characteristics are then related to those considered in Section 4.4, namely marginal means, marginal variances, correlation coefficients, and so on. As a statistically oriented application, we consider linear prediction of a random variable given another.

Since some of the statements made in Section 5.3, although plausible, are not easily proved except by considering conditional expectation from an abstract point of view, we devote Section 5.4 (Abstract Conditional Expectations) to doing just this. Although this last section is not needed for studying the rest of the book, it contains an elementary introduction to abstract conditional expectations which may be of use to the reader when he takes a more advanced course in probability.

5.1 CONDITIONAL DISTRIBUTIONS

Given a random variable X, the knowledge of its distribution P_X provides information concerning probabilities of various X-events. As we recall, for any Borel set A the probability of the X-event

$$X^{-1}(A) = \{\omega : X(\omega) \in A\}$$

is given by $P_X(A)$. Different choices of Borel sets A give rise to different X-events $X^{-1}(A)$, and the probabilities of these events are incorporated in a single probability measure P_X.

If two random variables X and Y both defined on the same probability space are considered, or, as some poeple prefer to say, if X and Y are jointly distributed, then the family of X-events and the family of Y-events naturally lead us to consider the conditional probability of a certain Y-event, say, $Y^{-1}(B)$, given a certain X-event, say $X^{-1}(A)$, where A and B are Borel sets in R. For example, if X and Y denote the height and the weight of an American adult, one may wish to consider the probability of an arbitrarily picked adult weighing more than 200 lbs given that he is not more than 5 ft tall. In this case $A = [0, 5]$, $B = [200, \infty)$ and we are interested in the conditional probability $P(Y \in B \,|\, X \in A)$ or more formally $P[Y^{-1}(B) \,|\, X^{-1}(A)]$. Now such a conditional probability is really nothing new since we have already encountered $P(B \,|\, A)$ in Section 2.1 where A and B were general events not necessarily describable in terms of X and Y. What will be new for us here is the set function arising from $P[Y^{-1}(B) \,|\, X^{-1}(A)]$ as we keep A fixed while allowing B to vary through the class of Borel sets. Since this set function assigns to each Borel set B a certain number $P[Y^{-1}(B) \,|\, X^{-1}(A)]$, we shall express this by writing

$$P_{Y|X \in A}(B) = P[Y^{-1}(B) \,|\, X^{-1}(A)]$$

The complete knowledge of the set function $P_{Y|X \in A}$, also denoted $P_{Y|X}(\cdot \,|\, A)$ whenever convenient, will provide information concerning probabilities of various Y-events given the particular X-event $X^{-1}(A)$. Of course the event $X^{-1}(A)$ must have a positive probability; otherwise, $P[Y^{-1}(B) \,|\, X^{-1}(A)]$ is not defined. However, as long as $P(X^{-1}(A)) > 0$, it is a simple exercise to check that the set function $P_{Y|X}(\cdot \,|\, A)$ has the following basic properties required of a probability distribution (or measure):

(i) For any Borel subset of B of the real line R, we have

$$0 \leqq P_{Y|X}(B|A) \quad \text{and} \quad P_{Y|X}(R|A) = 1$$

(ii) For any disjoint sequence of Borel sets $B_1, B_2, B_3, \ldots$, we have

$$P_{Y|X}\left(\bigcup_{i=1}^{\infty} B_i \,\middle|\, A\right) = \sum_{i=1}^{\infty} P_{Y|X}(B_i|A)$$

In other words $(R, \mathscr{B}, P_{Y|X \in A})$ is a probability space for each Borel set A such that $P(X \in A) > 0$. We now state formally the following definition.

5.1.1 DEFINITION

Given two jointly distributed random variables X and Y, for each Borel set A such that $P(X \in A) > 0$, we define the *conditional (probability) distribution* of Y given $X \in A$ by

$$P_{Y|X \in A}(B) = P(Y \in B \,|\, X \in A)$$
$$= \frac{P(X \in A, \, Y \in B)}{P(X \in A)}$$
$$= \frac{P_{X,Y}(A \times B)}{P_X(A)}$$

for each Borel set $B \in \mathscr{B}$.

Our definition of conditional distribution $P_{Y|X \in A}$ is somewhat lacking in generality, since A has to be such that $P(X \in A) > 0$. For example, $P_{Y|X=x}$ is not defined unless x happens to be one of the discrete points of X. However, if (X, Y) happens to be absolutely continuous, we may succeed in arriving at a meaningful definition of $P_{Y|X=x}$. We shall do this after introducing conditional density function $f_{Y|X=x}$ in the next section (see Definition 5.2.1).

5.1.1 PROPOSITION

Let a random vector $\mathbf{X} = (X, Y)$ have a discrete distribution with discrete points $x_i = (x_i, y_i)$ and the corresponding probability mass $P_{\mathbf{X}}(x_i) = p_i > 0$ for $i = 1, 2, 3, \ldots$, then for each x such that $x = x_i$ for some i the conditional distribution of Y given $X = x$ is discrete with discrete points y_j for all j such that $x_j = x$.

PROOF

Note first that for x such that $x = x_i$ for some i we have $P_X(x) > P_{\mathbf{X}}(x_i) > 0$. Now for each y_j with j such that $x_j = x$ we have by Definition 5.1.1:

$$P_Y(y_j \,|\, X = x) = \frac{P_{X,Y}[(x_j, y_j)]}{P_X(x)} > 0$$

Hence y_j is a discrete point for the conditional distribution $P_Y(\cdot \,|\, X = x)$. Taking the totality of all y_j with j such that $x_j = x$ we have

$$\sum_j P_Y(y_j \,|\, X = x) = \frac{\sum_j P_{X,Y}[(x_j, y_j)]}{P_X(x)} = \frac{P_X(x)}{P_X(x)} = 1$$

In other words the points y_j account for the entire probability mass 1. This completes the proof.

Given the joint distribution of (X, Y) the best way to visualize the conditional distribution $P_{Y|X \in A}$ is to look at the probability mass sitting on the vertical strip $A \times R_Y$ and project it horizontally onto the vertical axis R_Y. For example, if X and Y have the joint distribution sketched in Fig. 5.1.1, then $P_{Y|X \in A_1}$ with $A_1 = [0, 1]$ is a distribution centered around $Y = 0.5$, while $P_{Y|X \in A_2}$ with $A_2 = [1, 2]$ is a distribution centered somewhat around $Y = 1$. Incidentally, after projecting the mass onto R_Y we must remember to scale it up by the factor $1/P(X \in A) \ge 1$, since according to Definition 5.1.1 $P_{Y|X \in A}(B)$ is equal to the projected amount $P_{X,Y}(A \times B)$ divided by $P_X(A)$. Note also $P_{Y|X \in A}$ cannot be visualized if no mass is found in the vertical strip $A \times R_Y$, but then $P(X \in A) = 0$ in this case and we never attempted to define $P_{Y|X \in A}$ for such an A in the first place. In the following example we consider a specific conditional distribution which turns out to be describable in terms of a density function, conveniently denoted as $f_{Y|X \in A}(y)$ or $f_{Y|X}(y\,|\,A)$.

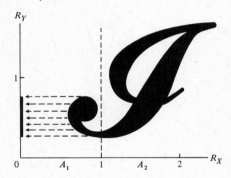

Figure 5.1.1

Example 1 Let X and Y have a continuous joint distribution over the unit square in the corner of the first quadrant of $R_X \times R_Y$ (see Fig. 5.1.2). If their joint density function is given by

$$f_{X,Y}(x, y) = 4xy$$

determine the conditional distribution $P_{Y|X}(\cdot \,|\, A)$ where A is the interval $[0, \frac{1}{2}]$.

Solution If X is confined in A, the range of Y will still be $[0, 1]$. A good way of determining any distribution (conditional or otherwise) is to find its distribution function. In the present case we shall calculate

$$F_{Y|X}(y\,|\,A) = P_{Y|X}[(-\infty, y]\,|\,A]$$

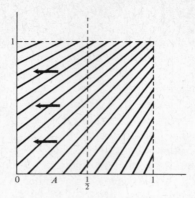

Figure 5.1.2

for any real y between 0 and 1. For $y \geq 1$, we have clearly $F_{Y|X}(y|A) = 1$; likewise for $y \leq 0$ we have $F_{Y|X}(y|A) = 0$. Now

$$
\begin{aligned}
F_{Y|X}(y|A) &= P(Y \leq y \,|\, 0 \leq X \leq \tfrac{1}{2}) \\
&= \frac{P(Y \leq y, \, 0 \leq X \leq \tfrac{1}{2})}{P(0 \leq X \leq \tfrac{1}{2})} \\
&= \frac{\int_0^{1/2} \int_0^y 4tu \, du \, dt}{\int_0^{1/2} \int_0^1 4tu \, du \, dt} \\
&= \frac{\tfrac{1}{4}y^2}{\tfrac{1}{4}} = y^2
\end{aligned}
$$

Differentiating $F_{Y|X \in A}(y)$ with respect to y, we obtain

$$f_{Y|X \in A}(y) = 2y$$

In other words more mass will be found toward $Y = 1$ after the projection.

As a special case of conditional distribution we have what is known as a marginal distribution. If in $P_{Y|X \in A}$ we let $A = R_X$, then for any Borel set B in R_Y we have simply

$$
\begin{aligned}
P_{Y|X \in R}(B) &= P(Y \in B \,|\, X \in R) \\
&= P(Y \in B) = P_Y(B)
\end{aligned}
$$

Thus the distribution P_Y can be visualized by projecting the probability mass in the vertical strip $R \times R$ (in other words the probability mass in the entire plane) onto the vertical axis R_Y. It is thus in reference to the joint distribution $P_{X, Y}$ that we speak of P_Y as the *marginal* distribution of Y. Reversing the roles of X and Y, we formulate the following definition.

5.1.2 DEFINITION

Given the joint distribution $P_{X, Y}$, by the *marginal distribution* of X we mean the conditional distribution $P_{X|Y \in R}$. Likewise the conditional distribution $P_{Y|X \in R}$ is called the *marginal distribution* of Y.

We note that aside from the context in which it arises the marginal distribution of X is no different from the distribution of X since $P_{X|Y \in R} = P_X$.

We now derive the concept of stochastic independence of two jointly distributed random variables within the context of conditional distributions just as we derived the concept of stochastic independence of two events within the context of conditional probabilities. Let X and Y be jointly distributed, then various conditional distributions $P_{Y|X \in A}$ may be considered simply by varying A. These conditional distributions are in general different depending on the choices of A, but if it should so happen that $P_{Y|X \in A}$ remains unchanged as a probability measure no matter what the choice of A is (in particular the whole R), then evidently any information regarding X provides no distinct information on Y, and we must regard Y as stochastically independent from X. Accordingly we consider the following definition.

5.1.3 DEFINITION
Given jointly distributed random variables X and Y, we say Y is stochastically independent from X if

$$P_{Y|X \in A} = P_Y$$

for all Borel sets A.

In other words for any Borel sets A and B we have

$$P_{Y|X \in A}(B) = P_Y(B)$$

or

$$\frac{P(X \in A, \ Y \in B)}{P(X \in A)} = P(Y \in B)$$

This last equation can be manipulated into

$$P(X \in A, \ Y \in B) = P(X \in A)P(Y \in B) \tag{1}$$

and

$$\frac{P(X \in A, \ Y \in B)}{P(Y \in B)} = P(X \in A)$$

so that we see $P_{X|Y \in B}(A) = P_X(A)$ for any A and for any B, thus

$$P_{X|Y \in B} = P_X$$

for all Borel sets B, meaning that X is independent of Y.

Stochastic independence of random variables is thus a symmetric relation and taking equation (1) above as the central condition, we formulate the following definition.

5.1.4 DEFINITION
Jointly distributed random variables X and Y are said to be (stochastically) independent if

$$P(X \in A, \ Y \in B) = P(X \in A)P(Y \in B) \tag{1}$$

for any Borel sets A and B, that is, an X-event $X^{-1}(A)$ and a Y-event $Y^{-1}(B)$ are always independent (thus our definition here is consistent with the earlier Definition 3.4.5).

With mere change in notation (1) may also be written as

$$P_{X,Y}(A \times B) = P_X(A) \cdot P_Y(B) \tag{2}$$

This equation is particularly significant since it says that the amount of probability mass on a "Borel rectangle" $A \times B$ is unequivocally determined by $P_X(A)$ and $P_Y(B)$. But since the amount of probability mass $P_{X,Y}(Q)$ on any Borel set Q is uniquely determined by the way the mass is distributed over the Borel rectangles (proved in measure theory), we see that the joint distribution $P_{X,Y}$ is completely determined by the marginal distributions P_X and P_Y if X and Y are independent. In this case one may wish to call $P_{X,Y}$ the *product distribution* of P_X and P_Y.

Returning to the concept of conditional distribution, we stress once again that $P_{Y|X=x}$ is not defined unless $P(X = x) > 0$. However, within the context of an actual experiment $P_{Y|X=x}$ may acquire an intuitive meaning even though $P_{Y|X=x}$ is not formally defined. Consider, for instance, the following example.

Example 2 A real number X is picked out from the real line according to the density function $f(x)$. If X takes the value x, a coin with probability $0 < p(x) < 1$ of falling heads is tossed. In this experiment what is the probability of the coin falling heads? That is, if we let Y be the number of heads (either 0 or 1), what is $P_Y(1)$?

In this experiment the "intuitive" conditional probability of $Y = 1$ given $X = x$ is $p(x)$ although the formal $P_{Y|X=x}$, let alone $P_{Y|X=x}(1)$, is undefined. In a situation like this, we shall write

$$\tilde{P}_{Y|X=x}(1) = p(x)$$

and call $\tilde{P}_{Y|X=x}$ the *preconditional probability distribution* of Y given $X = x$. In the present case, $\tilde{P}_{Y|X=x}$ is a two-point distribution for each x with $\tilde{P}_{Y|X=x}(1) = p(x)$ and $\tilde{P}_{Y|X=x}(0) = 1 - p(x)$. The justification for our considering the preconditional (probability) distribution $\tilde{P}_{Y|X=x}$ is that it enables us to set up a joint distribution $P_{X,Y}$ in a reasonable way, and out of this joint distribution we can determine the probabilities of events concerning Y. Specifically, we have the following useful proposition.

5.1.2 PROPOSITION

Let X be distributed according to the density function $f(x)$, and suppose $\tilde{P}_{Y|x}$ is a preconditional distribution for each x, that is, $\tilde{P}_{Y|x}$ is a probability measure for each x, such that for each Borel set B, $\tilde{P}_{Y|x}(B)$ is a piecewise continuous function of x, then there exists a unique probability distribution $P_{X,Y}$ such that

$$P_{X,Y}(A \times B) = \int_A \tilde{P}_{Y|x}(B) f(x)\, dx$$

In other words, $\tilde{P}_{Y|x}$ together with $f(x)$ "generates" a certain $P_{X,Y}$ that assigns to $A \times B$ the amount of probability mass that appears to us perfectly reasonable. In fact, the integral $\int_A \tilde{P}_{Y|x}(B) f(x)\, dx$ is the limit of the approximating sum $\sum_i \tilde{P}_{Y|x_i}(B) f(x_i)\, \Delta_i x$ in which $f(x_i)\, \Delta_i x$ is roughly the probability that X takes values in $\Delta_i x$ (see Fig. 5.1.3), and $\tilde{P}_{Y|x_i}(B)$ is the "intuitive" conditional probability of $Y \in B$ given $X = x_i$ so that in view of the partition formula (Proposition 2.2.2) the approximating sum above should be close to what we should consider as the probability of $(X, Y) \in A \times B$. The formal proof of Proposition 5.1.2 is a measure-theoretic manipulation involving the extension principle mentioned in Section 2.4. We now apply this proposition in answering the question posed in Example 2 of this section. Assuming $p(x)$ is a piecewise continuous function, we have according to Proposition 5.1.2

$$P_Y(1) = P_{X,Y}(R \times \{1\}) = \int_{-\infty}^{\infty} \tilde{P}_{Y|x}(1) f(x)\, dx$$

$$= \int_{-\infty}^{\infty} p(x) f(x)\, dx$$

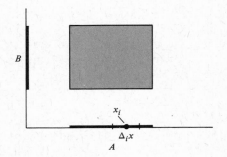

B

x_i

$\Delta_i x$

Figure 5.1.3 A

Example 3 A real number X is chosen from the interval $[0, 1]$ with uniform density. If $X = x$, a coin with probability x of falling heads is tossed repeatedly n times. Find the probability that the number of heads Y is equal to k for $k = 0, 1, 2, \ldots, n$.

Solution In this experiment we have the preconditional distribution $\tilde{P}_{Y|x}(k) = C_k^n x^k (1 - x)^{n-k}$. Hence, according to Proposition 5.1.2 we have

$$P_Y(k) = P_{X,Y}([0, 1] \times \{k\})$$

$$= \int_0^1 \tilde{P}_{Y|x}(k) \cdot 1\, dx$$

$$= \int_0^1 C_k^n x^k (1 - x)^{n-k}\, dx$$

To evaluate the integral we consider the so-called *beta function* defined by

$$B(r, s) = \int_0^1 x^{r-1}(1 - x)^{s-1} \, dx \qquad r, s > 0$$

together with its relation to the gamma function (cf. Definition 3.2.2 and Problem 5.1.8)

$$B(r, s) = \frac{\Gamma(r)\Gamma(s)}{\Gamma(r + s)}$$

Now then

$$
\begin{aligned}
P_Y(k) &= C_k^n B(k + 1, n - k + 1) \\
&= C_k^n \frac{\Gamma(k + 1)\Gamma(n - k + 1)}{\Gamma(n + 2)} \\
&= \frac{n!}{(n - k)!k!} \frac{k!(n - k)!}{(n + 1)!} \\
&= \frac{1}{n + 1}
\end{aligned}
$$

Thus the events of $0, 1, 2, \ldots, n$ heads are all equally likely.

Example 4 A real number X is chosen from the interval $(0, 1)$ with the density $f(x) = 3x^2$. If $X = x$, a real number Y is chosen from the interval $(0, x)$ with a uniform density. Find the probability of $X + Y < 1$.

Solution According to Proposition 5.1.2, there is a joint distribution $P_{X, Y}$ arising from the preconditional distribution $\tilde{P}_{Y|x}$, which is a uniform distribution on the interval $(0, x)$ for each $0 < x < 1$. The required probability is represented by the amount of probability mass in the shaded region Q in Fig. 5.1.4. Although we do not have a formula to calculate $P_{X, Y}(Q)$ when Q is not of the form $A \times B$,

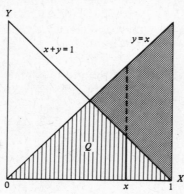

Figure 5.1.4

we can circumvent this difficulty as follows. Imagine a transformation of the plane in which the x-coordinate of each point remains fixed while the y-coordinate changes in such a way that Q is transformed into a region of the form $A \times B$ where $A = (0, 1)$ and $B = (0, \pi)$, say. Then we may set

$$\tilde{P}_{Y'|x}(B) = \begin{cases} 1 & \text{for} \quad 0 < x < \tfrac{1}{2} \\ \dfrac{1 - x}{x} & \text{for} \quad \tfrac{1}{2} \leqq x < \tfrac{1}{2} \end{cases}$$

Consequently, by Proposition 5.1.2 we have

$$P_{X,Y}(Q) = P_{X,Y'}(A \times B)$$

$$= \int_0^1 \tilde{P}_{Y'|x}(B)(3x^2)\, dx$$

$$= \int_0^{1/2} 1(3x^2)\, dx + \int_{1/2}^1 \frac{1 - x}{x}(3x^2)\, dx = \tfrac{3}{8}$$

Although Proposition 5.1.2 is useful as it is, it can be generalized by allowing X to be not necessarily absolutely continuous. We will state, but not prove, the following theorem.

5.1.1 THEOREM

Let X be a random variable having the distribution P_X. For each value x of X let there be given a distribution $\tilde{P}_{Y|x}$ such that for each Borel set B, $\tilde{P}_{Y|x}(B)$ is a Borel function of x, then there exists a unique distribution $P_{X,Y}$ such that

$$P_{X,Y}(A \times B) = \int_A \tilde{P}_{Y|x}(B)\, P_X(dx)$$

Furthermore, $P_{X,Y}$ is absolutely continuous if P_X and $\tilde{P}_{Y|x}$ for each x are absolutely continuous.

Two special comments can be made in connection with this theorem:
(1) If $\tilde{P}_{Y|x}(B)$ happens to be constant over A and if this constant is denoted by $\tilde{P}_Y(B)$, then

$$P_{X,Y}(A \times B) = \int_A \tilde{P}_Y(B)\, P_X(dx) = P_X(A)\tilde{P}_Y(B)$$

(2) If P_X and $\tilde{P}_{Y|x}$ for each x have density functions $f_X(x)$ and $\tilde{f}_{Y|x}(y)$, respectively, then

$$P_{X,Y}(A \times B) = \int_A \left[\int_B \tilde{f}_{Y|x}(y)\, dy \right] f_X(x)\, dx$$

$$= \int_{A \times B} f_X(x)\tilde{f}_{Y|x}(y)\, d(x, y)$$

so that $f_{X,Y}(x, y) = f_X(x)\tilde{f}_{Y|x}(y)$.

In Section 5.2 we shall reconcile our intuitive notion of preconditional density function $\tilde{f}_{Y|x}$ with the more formal notion of conditional density function $f_{Y|x}$.

Example 5 Let X be distributed on $[0, \infty)$ according to the density function $f_X(x) = e^{-x}$, and suppose the preconditional distribution of Y given $X = x > 0$ is such that $\tilde{P}_{Y|x}(-\infty, x] = 0$ and $\tilde{P}_{Y|x}(x, y] = 1 - e^{x-y}$. Determine the joint distribution of X and Y.

Solution Since P_X is absolutely continuous with the density function $f_X(x) = e^{-x}$, and $\tilde{P}_{Y|x}$ for each x is also absolutely continuous with the density function $\tilde{f}_{Y|x}(y) = 0$ for $y < x$ and $\tilde{f}_{Y|x}(y) = d/dy(1 - e^{x-y}) = e^{x-y}$ for $y \geq x$, in view of comment (2) above we have an absolutely continuous $P_{X,Y}$ with the density function

$$f_{X,Y}(x, y) = f_X(x)\tilde{f}_{Y|x}(y) = e^{-x}e^{x-y} = e^{-y}$$

for $y \geq x$.

5.1 PROBLEMS

1. Let the random vector (X, Y) have the discrete points $(1, 4)$, $(2, 1)$, $(3, 3)$, $(4, 1)$, $(4, 2)$ with corresponding probabilities $\frac{3}{10}, \frac{2}{10}, \frac{1}{10}, \frac{3}{10}, \frac{1}{10}$. Determine the conditional distribution of Y given $X \in A = [2, 4]$, also the conditional distribution of X given $Y = 1$.

2. If X and Y are jointly distributed in the unit square $0 \leq x \leq 1, 0 \leq y \leq 1$ according to the density function

$$f_{X, Y}(x, y) = x + y$$

determine the conditional distribution $P_{Y|X \in A}$ where $A = [0, 1/n]$. What happens when n tends to ∞?

3. Let (X, Y) be such that $\frac{1}{2}$ of the probability mass is uniformly distributed over the unit square $0 \leq x \leq 1, 0 \leq y \leq 1$, $\frac{1}{4}$ of the mass is uniformly distributed over the line segment connecting $(0, \frac{1}{2})$ and $(1, \frac{1}{2})$, $\frac{1}{8}$ of the mass is uniformly distributed over the line segment connecting $(0, \frac{1}{4})$ and $(1, 1)$, and finally $\frac{1}{8}$ of the mass is concentrated at the point $(1, 1)$. Sketch the distribution functions for the marginal distributions of X and Y.

4. Let (X, Y) be evenly distributed on the points $(1, 1)$ and $(-1, -1)$. If (U, V) has identical marginal distributions as (X, Y) and if U and V are independent, determine the distribution of (U, V).

5. If (X, Y) has a six-point discrete distribution at $(1, 1)$, $(1, 2)$, $(1, 3)$, $(2, 1)$, $(2, 2)$, $(2, 3)$ with the corresponding probabilities $\frac{1}{6}, \frac{1}{9}, \frac{1}{18}, \frac{1}{3}, \alpha, \beta$, what must α and β be so that X and Y will be independent?

6. Which of the following distributions for (X, Y) make X and Y independent?
a. Uniform distribution over a rectangle with sides parallel to the coordinate axes R_X and R_Y.
b. Uniform distribution over the boundary of a rectangle in (a).
c. Uniform distribution over the line segment connecting $(1, 1)$ and the origin.
d. Even distribution over the three vertexes of a triangle.

7. A real number x is chosen from the interval $[0, 1]$ with uniform density; a coin with probability x^2 of falling heads is then tossed. What is the probability of no heads in this experiment?

8. Gamma and beta functions. Let the gamma function and the beta function, respectively, be defined by

$$\Gamma(r) = \int_0^\infty t^{r-1} e^{-t} \, dt \qquad r > 0$$

$$B(r, s) = \int_0^1 t^{r-1} (1 - t)^{s-1} \, dt \qquad r, s > 0$$

Show $B(r, s) = \Gamma(r)\Gamma(s)/\Gamma(r + s)$ by the following steps:
a. Let $t = x^2$ in $\Gamma(r)$ and $t = y^2$ in $\Gamma(s)$ to obtain

$$\Gamma(r)\Gamma(s) = 4 \int_0^\infty \int_0^\infty x^{2r-1} y^{2s-1} e^{-(x^2+y^2)} \, dx \, dy$$

b. Express $\Gamma(r)\Gamma(s)$ above in terms of the polar coordinates $x = \rho \cos \theta$, $y = \rho \sin \theta$, to obtain

$$\Gamma(r)\Gamma(s) = 4 \int_0^{\pi/2} \int_0^\infty (\cos \theta)^{2r-1} (\sin \theta)^{2s-1} e^{-\rho^2} \rho^{2r+2s-1} \, d\rho \, d\theta$$

c. Set $\rho^2 = u$ in $\int_0^\infty e^{2r+2s-1} e^{-\rho^2} \, d\rho$ to obtain

$$\Gamma(r)\Gamma(s) = \int_0^{\pi/2} (\cos \theta)^{2r-1} (\sin \theta)^{2s-1} [2\Gamma(r + s)] \, d\theta$$

d. Let $u = \cos^2 \theta$ in the integral to obtain

$$\Gamma(r)\Gamma(s)/\Gamma(r + s) = -\int_1^0 u^{r-1} (1 - u)^{s-1} \, du = B(r, s)$$

9. A real number X is chosen from the interval $[0, \infty)$ with the density $f(x) = e^{-x}$. A real number Y is chosen from the interval $[0, 1]$ with uniform density. Find the probability for $X \geq Y$.

5.2 MARGINAL AND CONDITIONAL DENSITY FUNCTIONS

We specialize earlier results in the preceding section to the absolutely continuous cases. First of all, since the joint distribution $P_{X,Y}$ uniquely determines the marginal distributions P_X and P_Y, if the joint density function $f_{X,Y}$ is given we ought to be able to calculate the marginal density functions f_X and f_Y. Accordingly we consider the following proposition.

5.2.1 PROPOSITION

If X and Y have an absolutely continuous joint distribution with density function $f_{X,Y}$, then X and Y are each absolutely continuous with density functions calculated as follows:

$$f_X(x) = \int_{-\infty}^{\infty} f_{X,Y}(x, y) \, dy$$

$$f_Y(y) = \int_{-\infty}^{\infty} f_{X,Y}(x, y) \, dx$$

PROOF

For any Borel set A we have

$$P_X(A) = P_{X,Y}(A \times R)$$

$$= \iint_{A \times R} f_{X,Y}(x, y) \, d(x, y)$$

$$= \int_A \left(\int_{-\infty}^{\infty} f_{X,Y}(x, y) \, dy \right) dx$$

In other words, $\int_{-\infty}^{\infty} f_{X,Y}(x, y) \, dy$ as a function in x plays the role of a density function for X. Likewise we see that $\int_{-\infty}^{\infty} f_{X,Y}(x, y) \, dx$ as a function in y plays the role of a density function for Y.

Example 1 If X and Y have a uniform distribution over the disk of radius 1 around the origin of R^2, what density functions must X and Y have separately?

Solution The joint density function of X and Y is given by $f_{X,Y}(x, y) = 1/\pi$ for all (x, y) with $x^2 + y^2 \leq 1$. Therefore we have

$$f_X(x) = \int_{-\infty}^{\infty} f_{X,Y}(x, y) \, dy$$

$$= \int_{-\sqrt{1-x^2}}^{\sqrt{1-x^2}} \left(\frac{1}{\pi} \right) dy$$

$$= \frac{2}{\pi} \sqrt{1 - x^2} \qquad \text{for} \quad -1 \leq x \leq 1$$

Likewise

$$f_Y(y) = \frac{2}{\pi}\sqrt{1 - y^2} \qquad \text{for } -1 \leq y \leq 1$$

Next, we recall that the joint distribution of X and Y is uniquely determined by the marginal distributions of X and Y if X and Y are independent (cf. remarks following Definition 5.1.4). Thus if X and Y are independent and f_X and f_Y are known, we ought to be able to calculate the joint density function $f_{X,Y}$. This leads us to the following proposition.

5.2.2 PROPOSITION

The random vector (X, Y) has an absolutely continuous distribution if X and Y are independent and are both absolutely continuous. The density function of (X, Y) is obtained from the *product formula*:

$$f_{X,Y}(x, y) = f_X(x)f_Y(y)$$

PROOF
Since X and Y are independent, we recall

$$P_{X,Y}(A \times B) = P_X(A) \cdot P_Y(B)$$

for any Borel sets A and B. But

$$P_X(A) \cdot P_Y(B) = \left(\int_A f_X(x)\,dx\right)\left(\int_B f_Y(y)\,dy\right)$$

$$= \iint_{A \times B} [f_X(x)f_Y(y)]\,d(x, y)$$

In other words, $f_X(x)f_Y(y)$ as a function of two variables plays the role of a density function $P_{X,Y}$. Strictly speaking we should show

$$P_{X,Y}(Q) = \iint_Q f_X(x)f_Y(y)\,d(x, y)$$

for any Borel set Q in R^2 not just for $A \times B$. This is actually shown in measure theory by using the result we obtained for $A \times B$.

Example 2 An executive arrives at his office with "equal likelihood" at any time between 8 A.M. and 12 noon. His secretary arrives at the office with "equal likelihood" at any time between 7 A.M. and 9 A.M. Assuming that their arrival times are independent (cf. Problem 5.2.5), determine the probability of their arriving within 5 minutes of each other.

Solution Let X and Y be their arrival times. Assuming that X and Y are uniformly distributed over the intervals [8, 12] and [7, 9], respectively, we have

$$f_X(x) = \tfrac{1}{4} \quad \text{for} \quad 8 \leq x \leq 12$$
$$f_Y(y) = \tfrac{1}{2} \quad \text{for} \quad 7 \leq y \leq 9$$

Since X and Y are independent, by Proposition 5.2.2, (X, Y) is absolutely continuous with the density function

$$f_{X,Y}(x, y) = \tfrac{1}{4} \cdot \tfrac{1}{2} = \tfrac{1}{8} \quad \text{for} \quad 8 \leq x \leq 12,\, 7 \leq y \leq 9$$

Now the event that the executive and the secretary arrive within 5 minutes of each other is represented by a certain subset B in R^2 (see Fig. 5.2.1). Consequently we obtain

$$P_{X,Y}(B) = \iint_B f_{X,Y}(x, y)\, d(x, y)$$

$$= \tfrac{1}{8} \text{ Area } (B)$$

but since

$$\text{Area } (B) = \tfrac{1}{2}[(\tfrac{13}{12})^2 - (\tfrac{11}{12})^2] = \tfrac{1}{6}$$

we obtain

$$P_{X,Y}(B) = \tfrac{1}{48} \doteq 0.021$$

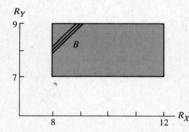

Figure 5.2.1

Example 3 (A Derivation of Normal Distribution) A dart is thrown at a large coordinate plane with the aim to hit the origin. Let X and Y be the x and y coordinates of the point hit by the dart. The following assumptions are made:
  (i) X and Y are independent.
 (ii) X and Y are both absolutely continuous with identical densities.
(iii) X and Y have a spherical joint distribution (cf. Problem 3.4.3).

Determine the density functions f_X, f_Y, and $f_{X,Y}$ up to a few unspecified parameters.

Solution Since X and Y are independent and identically distributed

$$f_{X,Y}(x, y) = f_X(x)f_Y(y) = f_X(x)f_X(y)$$

Writing g for $f_{X,Y}$ and f for f_X, we reduce this to

$$g(x, y) = f(x)f(y) \tag{1}$$

Since g is spherical, we have in particular

$$g(x, y) = g(r, 0) \tag{2}$$

where $r^2 = x^2 + y^2$ (see Fig. 5.2.2). But

$$g(r, 0) = f(r)f(0)$$

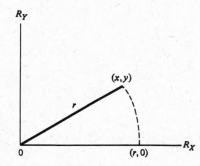

Figure 5.2.2

Hence in view of (1) and (2) we have

$$f(x)f(y) = f(r)f(0)$$

To solve this equation for f we take the logarithm

$$\log f(x) + \log f(y) = \log f(r) + \log f(0)$$

then writing $\phi(x)$ for $\log f(x)$, and differentiating both sides with respect to x, we obtain

$$\phi'(x) = \phi'(r)\frac{\partial r}{\partial x}$$

But $\partial r/\partial x = x/r$ by implicit differentiation of $r^2 = x^2 + y^2$. Hence

$$\frac{\phi'(x)}{x} = \frac{\phi'(r)}{r}$$

If we keep x fixed and vary y, then r varies and yet $\phi'(r)/r$ remains equal to $\phi'(x)/x$ and is unchanged. In other words, $\phi'(r)/r$ is a constant function. This constant has to be negative, for otherwise $\phi'(r) \geq 0$ for $r > 0$ and $\phi(r)$ would be monotone increasing for $r > 0$ and so would be $f(r) = e^{\phi(r)}$, and thus

$\int_0^\infty f(r)\,dr = \infty$, contradicting the fact that f is a density function. Accordingly we set, after replacing the dummy variable r by x,

$$\frac{\phi'(x)}{x} = -\alpha \qquad \text{for} \quad \alpha > 0$$

From $\phi'(x) = -\alpha x$ now follows

$$\phi(x) = -\tfrac{1}{2}\alpha x^2 + c$$

and so

$$f(x) = \beta e^{-(1/2)\alpha x^2}$$

From $1 = \int_{-\infty}^\infty \beta e^{-(1/2)\alpha x^2}\,dx$ we determine

$$\beta = \frac{\sqrt{\alpha}}{\sqrt{2\pi}}$$

Finally letting $\sqrt{\alpha} = 1/\sigma$ we arrive at

$$f_X(x) = \frac{1}{\sigma\sqrt{2\pi}} \exp\left[-\frac{1}{2}\left(\frac{x-0}{\sigma}\right)^2\right]$$

In other words X (and hence also Y) has $(0, \sigma)$ normal distribution. As for $f_{X,Y}$ we have

$$f_{X,Y}(x, y) = f_X(x)f_Y(y) = \frac{1}{2\pi\sigma^2} \exp\left[-\frac{1}{2\sigma^2}(x^2 + y^2)\right]$$

We have found that $f_{X,Y}(x, y) = f_X(x)f_Y(y)$ if X and Y are independent, but what if X and Y are not independent? Can we still factor $f_{X,Y}(x, y)$ somehow so that the resulting product formula generalizes the above formula for independent X and Y? We shall do this with the aid of conditional density functions to be introduced via Proposition 5.2.3 and Definition 5.2.1 below. The resulting formula is as follows:

$$f_{X,Y}(x, y) = f_X(x)f_Y(y\,|\,x)$$

5.2.3 PROPOSITION
If (X, Y) is absolutely continuous, then the conditional distribution $P_{Y|X \in A}$ is absolutely continuous for each Borel set A.

PROOF
For each Borel set B we have

$$P_{Y|X \in A}(B) = \frac{P(X \in A, Y \in B)}{P(X \in A)}$$

$$= \frac{\int_B \int_A f_{X,Y}(x, y)\,dx\,dy}{\int_A f_X(x)\,dx}$$

$$= \int_B \left[\frac{\int_A f_{X,Y}(x, y)\,dx}{\int_A f_X(x)\,dx}\right] dy$$

Thus $P_{Y|X \in A}$ is absolutely continuous with the density function

$$f_{Y|X \in A}(y) = \frac{\int_A f_{X,Y}(x, y)\, dx}{\int_A f_X(x)\, dx}$$

And now we can let A shrink to a fixed number x_0 to arrive at the conditional density function of Y at $X = x_0$, denoted briefly $f_{Y|x_0}$.

5.2.1 DEFINITION

Given an absolutely continuous (X, Y) we define the *conditional density function* of Y given $X = x_0$ by

$$f_{Y|x_0}(y) = \lim_{0 < h \to 0} f_{Y|X \in [x_0, x_0+h]}(y)$$

$$= \lim_{0 < h \to 0} f_{Y|X \in [x_0-h, x_0]}(y)$$

for all x_0 and y such that the two limits exist and are equal. The probability distribution determined by the density function $f_{Y|x_0}$ will be called the *conditional distribution* of Y given $X = x_0$, denoted $P_{Y|x_0}$.

Although $f_{Y|x_0}(y)$ is not defined in general for all (x_0, y), for some important special cases it is (see Corollary 1 of this section).

5.2.4 PROPOSITION

If (X, Y) is absolutely continuous, then

$$f_{Y|x_0}(y) = \frac{f_{X,Y}(x_0, y)}{f_X(x_0)}$$

for each (x_0, y) such that (1) $f_X(x_0) > 0$, (2) $f_X(x)$ is continuous in some neighborhood of x_0, and (3) $f_{X,Y}(x, y)$ is continuous in x in some neighborhood of x_0 for every y.

The three conditions we have to impose on (x_0, y) are all automatically satisfied if the joint density function $f_{X,Y}(x, y)$ is continuous and positive everywhere (see Example 5 of this section).

PROOF

By the result of the preceding proposition and the mean value theorem of integral calculus, we have

$$f_{Y|X \in [x_0, x_0+h]}(y) = \frac{\int_{x_0}^{x_0+h} f_{X,Y}(x, y)\, dx}{\int_{x_0}^{x_0+h} f_X(x)\, dx}$$

$$= \frac{f_{X,Y}(\bar{x}, y)h}{f_X(x^*)h}$$

where $\bar{x}$ and x^* are some points in the interval $[x_0, x_0 + h]$. As h tends to 0,

$\bar{x}$ and x^* both tend to x_0, and hence by the continuity of f_X and $f_{X,Y}$ as assumed, we obtain

$$\lim_{h \to 0} f_{Y|X \in [x_0, x_0 + h]}(y) = \frac{f_{X,Y}(x_0, y)}{f_X(x_0)}$$

Likewise we could have shown

$$\lim_{h \to 0} f_{Y|X \in [x_0 - h, x_0]}(y) = \frac{f_{X,Y}(x_0, y)}{f_X(x_0)}$$

Consequently we have

$$f_{Y|x}(y) = \frac{f_{X,Y}(x, y)}{f_X(x)}$$

or

$$f_{X,Y}(x, y) = f_X(x) f_Y(y|x)$$

This equation was mentioned just before Proposition 5.2.3. Note also its resemblance to $P(AB) = P(A)P(B|A)$, the product formula first mentioned in Section 2.1.

5.2.1 COROLLARY

If (X, Y) has an everywhere continuous and positive density function $f_{X,Y}(x, y)$, then the conditional density function $f_{Y|x}(y)$ is defined for all y for all x and in fact

$$f_{Y|x}(y) = \frac{f_{X,Y}(x, y)}{f_X(x)}$$

where $f_X(x)$ is the marginal density function of X.

A simple geometric interpretation of conditional density function $f_{Y|x}(y)$ is possible in view of the expression $f_{X,Y}(x, y)/f_X(x)$. We do this by cutting the surface of $f_{X,Y}(x, y)$ by an "x-plane" (see Fig. 5.2.3) and scaling the intersection curve up (or down) by $1/f_X(x)$. Thus $f_{Y|x}(y)$ can be taken to indicate what values of Y are more likely when $X = x$. Note also that $f_{Y|x}(y)$ is a nonnegative function and behaves like a density function. Indeed we have

$$\int_{-\infty}^{\infty} f_{Y|x}(y) \, dy = \frac{1}{f_X(x)} \int_{-\infty}^{\infty} f_{X,Y}(x, y) \, dy$$

$$= \frac{f_X(x)}{f_X(x)} = 1$$

If X and Y happen to be independent, then from $f_{X,Y}(x, y) = f_X(x) f_Y(y)$ follows

$$f_{Y|x}(y) = \frac{f_{X,Y}(x, y)}{f_X(x)} = f_Y(y)$$

so that all conditional density functions of Y given various x are equal. Geometrically this means that the cross sections of the surface $f_{X,Y}(x, y)$ by various "x-planes" are all alike, differing only by scaling factors.

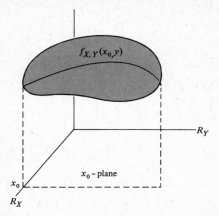

Figure 5.2.3

Strictly speaking, Corollary 5.2.1, which says that under certain special conditions we have $f_{Y|x}(y) = f_{X,Y}(x, y)/f_X(x)$, amounts to a particular realization of Definition 5.2.1. That is, if we confine ourselves to certain special cases, then $f_{X,Y}(x, y)/f_X(x)$ is as good a definition of conditional density function as Definition 5.2.1. In fact, it may even be regarded as a better definition since it enables us to formally determine $f_{Y|x}$ from $f_{X,Y}$. Now since the choice of a mathematical definition is largely a subjective matter restricted by no other criterion than that the chosen definition make good sense and work formally, it often happens that an initial definition is supplanted by another. In our case, the new definition is preferred on several counts: (1) the integral $\int_{-\infty}^{\infty} f_{X,Y}(x, y)/f_X(x) \, dy$ is obviously equal to 1; (2) the (formal) conditional density function $f_{Y|x}(y) = f_{X,Y}(x, y)/f_X(x)$ is consistent with the (intuitive) preconditional distribution $\tilde{P}_{Y|x}(B)$ introduced in the preceding section. To be more precise we state the following proposition.

5.2.5 PROPOSITION

Suppose X is distributed according to the density function $f_X(x)$ and the preconditional distribution $\tilde{P}_{Y|x}$ is such that $\tilde{P}_{Y|x}(B)$ is a piecewise continuous function of x for each Borel set B (cf. Proposition 5.1.2), and suppose further that the joint distribution $P_{X,Y}$ arising from f_X and $\tilde{P}_{Y|x}$ (again cf. Proposition 5.1.2) is absolutely continuous with density function $f_{X,Y}(x, y)$, then for almost all x (that is, excepting a set of probability 0) we have

$$\int_B \frac{f_{X,Y}(x, y)}{f_X(x)} \, dy = \tilde{P}_{Y|x}(B)$$

PROOF

We need only show that the function

$$\phi(x) = \int_B \frac{f_{X,Y}(x, y)}{f_X(x)} \, dy - \tilde{P}_{Y|x}(B)$$

$$= \frac{1}{f_X(x)} \left[\int_B f_{X, Y}(x, y) - \tilde{P}_{Y|x}(B) f_X(x) \right]$$

vanishes for almost all x. But then for any Borel set A we have

$$\int_A \left[\int_B f_{X, Y}(x, y) \, dy - \tilde{P}_{Y|x}(B) f_X(x) \right] dx$$

$$= \int_A \int_B f_{X, Y}(x, y) \, dy \, dx - \int_A \tilde{P}_{Y|x}(B) f_X(x) \, dx$$

$$= P_{X, Y}(A \times B) - P_{X, Y}(A \times B) = 0$$

in view of Proposition 5.1.2. This completes the proof.

We now discard Definition 5.2.1 and formally adopt the following definition.

5.2.2 DEFINITION

For an absolutely continuous (X, Y) with density function $f_{X, Y}(x, y)$ we define the conditional density function of Y given $X = x$ by

$$f_{Y|x}(y) = \frac{f_{X, Y}(x, y)}{f_X(x)}$$

for all x at which $f_X(x) > 0$.

As for x at which $f_X(x) = 0$, we may ignore such x since they are extremely rare. In fact we have the following proposition.

5.2.6 PROPOSITION

If (X, Y) has an absolutely continuous distribution, then $P\{x : f_X(x) = 0\} = 0$.

PROOF

Let $A = \{x : f_X(x) = 0\}$, then

$$P_X(A) = P_{X, Y}(A \times R)$$

$$= \int_A \int_{-\infty}^{\infty} f_{X, Y}(x, y) \, dy \, dx$$

$$= \int_A f_X(x) \, dx \qquad \text{(by Proposition 5.2.1)}$$

$$= 0$$

We must admit that it is quite unreasonable to insist on the definition of $f_{Y|x}$ for all x. For example, if X and Y are such that they never take values larger than, say, π, why should we bother considering, say, $f_{Y|4}$?

Example 4 Given $f_{X,Y}(x, y) = e^{-y}$ for $0 \le x \le y$ and $=0$ everywhere else, find the conditional density function $f_{Y|x}(y)$ then the conditional distribution function $F_{Y|x}(y) = P(Y \le y \mid X = x)$.

Solution By Definition 5.2.2 we have $f_{Y|x}(y) = f_{X,Y}(x, y)/f_X(x)$. But $f_X(x) = \int_x^\infty e^{-y} \, dy = e^{-x}$ for $x \ge 0$ and $=0$ elsewhere, therefore

$$f_{Y|x}(y) = \frac{e^{-y}}{e^{-x}} = e^{x-y} \qquad \text{for} \quad y \ge x$$

Now $F_{Y|x}(y) = 0$ for $y \le x$, and for $y \ge x$ we have

$$F_{Y|x}(y) = \int_x^y f_{Y|x}(u) \, du = \int_x^y e^x \cdot e^{-u} \, du = 1 - e^{x-y}$$

We conclude this section by a brief discussion of binormal (or bivariate normal) density function (see also Problems 3.6.18 and 5.3.7).

Example 5 If (X, Y) has the so-called binormal distribution specified by the density function

$$f(x, y) = \frac{1}{2\pi\sqrt{1 - \rho^2}} \exp\left[-\frac{1}{2(1 - \rho^2)} (x^2 - 2\rho xy + y^2) \right].$$

where ρ is some constant between -1 and 1, determine all conditional distributions by calculating the conditional density functions $f(y|x)$ and $f(x|y)$.

Solution Since $f(y|x) = f(x, y)/f(x)$, we must first determine the marginal density function $f(x)$. Now

$$f(x) = \int_{-\infty}^\infty f(x, y) \, dy$$

and after a tedious routine integration we obtain

$$f(x) = \frac{1}{\sqrt{2\pi}} e^{-x^2/2}$$

In other words the marginal distributions of X and likewise of Y are standard normal. Dividing $f(x, y)$ by $f(x)$ we obtain after simplification

$$f(y|x) = \frac{1}{\sqrt{2\pi}\sqrt{1 - \rho^2}} \exp\left[-\frac{1}{2}\left(\frac{y - \rho x}{\sqrt{1 - \rho^2}} \right)^2 \right]$$

In other words the conditional distribution of Y given $X = x$ is $(\rho x, \sqrt{1 - \rho^2})$ normal. Geometrically this means the cross sections of $f(x, y)$ by planes perpendicular to the R_X-axis are all alike except the scaling factors and shifting positions (see Fig. 5.2.4). Thus ρ has the effect of making the hump of the surface rounder (for ρ close to 0) or narrower (for ρ close to 1).

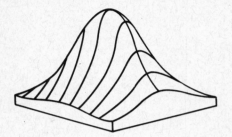

Figure 5.2.4

5.2 PROBLEMS

1. Let the random vector (X, Y) be uniformly distributed over the quadrilateral with vertexes at $(1, 0)$, $(0, 2)$, $(-3, 0)$, $(0, -4)$. Find the marginal density functions f_X and f_Y and sketch their graphs. Avoid integrations as much as possible.

2. Let (X, Y) be uniformly distributed over a rectangle with sides parallel to coordinate axes R_X and R_Y. Show that $P_{X, Y}$ is a product distribution by checking $P_{X, Y}(A \times B) = P_X(A)P_Y(B)$ for any intervals A in R_X and B in R_Y.

3 a. Given (X, Y) show that X and Y are independent if and only if (X, Y) has the density function in the form

$$f_{X, Y}(x, y) = g(x)h(y)$$

Hint: Show that $P_{Y|X \in A}(B)$ does not depend on A.

b. Given a binormally distributed (X, Y) (see Section 5.2, Example 5) show that X and Y are independent if and only if $\rho = 0$.

4. According to Proposition 5.2.1 the absolute continuity of (X, Y) guarantees the absolute continuity of X and of Y. Is the converse true? Hint: Consider (X, Y) uniformly distributed over the line segment connecting $(1, 1)$ and the origin.

5. If the arrival time X of an executive is uniformly distributed over the time interval $[8, 12]$ and the arrival time Y of his secretary is uniformly distributed over the time interval $[7, 9]$, show at least three joint distributions of (X, Y) which make X and Y not independent. Are these distributions absolutely continuous, singularly continuous, discrete?

6. Let (X, Y) be distributed according to the density function

$$f_{X, Y}(x, y) = xy + \tfrac{3}{2}x \qquad \text{for} \quad 0 \leq x \leq 1, 0 \leq y \leq 1$$

Find the conditional density function $f_{Y|x}$ for each x in $[0, 1]$.

7. Let X and Y be jointly distributed in the unit square with

$$f_{X, Y}(x, y) = xy + 2.25x^2 \quad \text{for} \quad 0 \leq x \leq 1, 0 \leq y \leq 1$$

a. Find the conditional density function $f_{Y|x}$ for all x in $[0, 1]$.

b. Use the conditional density function $f_{Y|x}$ for $x = \frac{1}{2}$ to approximate the conditional probability $P(\frac{1}{2} \leq Y \leq 1 | \frac{1}{2} \leq X \leq \frac{9}{16})$.

8. The formal analogy between the product formula

$$f(x, y) = f(x)f(y | x) \tag{1}$$

and the earlier product formula (Section 2.2)

$$P(AB) = P(A)P(B | A)$$

suggests further the following analogs of other earlier conditional probability formulas:

$$f(y) = \int_{-\infty}^{\infty} f(x)f(y | x)\, dx \quad \text{(partition formula)} \tag{2}$$

$$f(x | y) = \frac{f(x)f(y | x)}{\int_{-\infty}^{\infty} f(x)f(y | x)\, dx} \quad \text{(Bayes' formula)} \tag{3}$$

Prove (2) by using (1) and Proposition 5.2.1, then prove (3) by combining (2) and the dual of (1): $f(x, y) = f(y)f(x | y)$.

5.3 CONDITIONAL EXPECTED VALUES

Although mathematical expectation was defined for random variables, in essence it was defined for distributions of these random variables since as we recall

$$E(X) = \int_R x\, P_X(dx) = \begin{cases} \sum_i x_i p_i & \text{(discrete case)} \\ \int_{-\infty}^{\infty} x f_X(x)\, dx & \text{(density case)} \end{cases}$$

Therefore if a conditional distribution $P_{Y|X=x}$ is considered for the jointly distributed X and Y at $X = x$, we can also consider the mathematical expectation for the distribution $P_{Y|x}$. The resulting expected value $E(Y | x)$ is essentially the probability-weighted average of Y given $X = x$. We state formally the following definition.

5.3.1 DEFINITION

Given jointly distributed random variables X and Y the conditional expected value of Y given $X = x$ is defined by

$$E(Y | x) = \int_R y\, P_{Y|x}(dy) = \begin{cases} \sum_i y_i P_{Y|x}(y_i) & \text{(discrete case)} \\ \int_{-\infty}^{\infty} y f_{Y|x}(y)\, dy & \text{(density case)} \end{cases}$$

provided that the series and the integral are summable.

Note that our definition of $E(Y|x)$ presupposes the consideration of $P_{Y|x}$, which has not been defined for all cases. A more general definition of $E(Y|x)$ not relying on $P_{Y|x}$ will be given in the next section.

Example 1 A real number X is chosen from the interval $[0, 1]$ with uniform density. If $X = x$, a coin with probability x of falling heads is tossed n times. Let Y be the number of heads. Find $E(Y|x)$ for $0 \leqq x \leqq 1$.

Solution Since $P_{Y|x}(k) = C_k^n x^k (1 - x)^{n-k}$ for $k = 0, 1, 2, \ldots, n$, in other words, $P_{Y|x}$ is an (n, x) binomial distribution, its expected value is nx. Hence $E(Y|x) = nx$.

If $E(Y|x)$ is defined for every x, or over an interval of x, then the graph of $y = E(Y|x)$ shows how the conditional expected value of Y varies as x varies. This graph is known by its historic name of *regression curve*. The word "regression" is used in the sense of "going back," that is, the expected value $E(Y|x)$ goes back to (or depends on) the value of X.

Example 2 If X and Y are jointly distributed according to the binormal density function

$$f(x, y) = \frac{1}{2\pi\sqrt{1 - \rho^2}} \exp\left[-\frac{1}{2(1 - \rho^2)} (x^2 - 2\rho xy + y^2) \right]$$

determine $E(Y|x)$ for all x.

Solution $P_{Y|x}$ has a density function for each x obtained as

$$f(y|x) = \frac{f(x, y)}{f_1(x)}$$

where $f_1(x)$ is the marginal density function of X. According to Section 5.2, Example 5, we have after some calculations

$$f(y|x) = \frac{1}{2\pi\sqrt{1 - \rho^2}} \exp\left[-\frac{1}{2}\left(\frac{y - \rho x}{1 - \rho^2} \right)^2 \right]$$

Since this is exactly the density function of $(\rho x, \sqrt{1 - \rho^2})$ normal distribution with mean value equal to ρx, we see

$$E(Y|x) = \rho x$$

Incidentally, since $y = \rho x$ is the equation of a straight line with slope ρ, this result gives a geometric interpretation of ρ.

The two preceding examples are special cases of what is known among statisticians as linear regression. More generally we state the following definition.

5.3.2 DEFINITION

Given jointly distributed random variables X and Y we say that Y has a *linear regression* on X if the regression graph of Y on X is straight, that is:

$$E(Y|x) = \alpha + \beta x$$

for some constants α and β. The constants α and β are called the *regression coefficients* of Y on X.

Regression coefficients α and β describe roughly how Y depends on X; for example if $\beta > 1$ then for large values of X we expect Y to assume large values on the average. Now the correlation coefficient ρ of X and Y introduced in Section 4.4 also describe roughly how X and Y are related. Therefore, there should be some close relation between the regression coefficients and the correlation coefficient. We will now investigate this. First, however, we need develop a useful tool.

Recall that the conditional expected value $E(Y|x)$ depends on the value x, that is

$$E(Y|x) = g(x)$$

is a function of x. If g is a Borel function as it usually is, we may compose X and g to obtain a new random variable

$$g(X) = E(Y|X)$$

The question is: What do we obtain as the expected value of $g(X)$? That is:

$$E[g(X)] = E[E(Y|X)] = ?$$

We are essentially asking: What is the average of the conditional (or, say, "partial") averages of Y? As one might expect, this turns out to be exactly the average of Y. We formulate this as follows.

5.3.1 PROPOSITION

For jointly distributed random variables X and Y we have

$$E[E(Y|X)] = E(Y)$$

and likewise

$$E[E(X|Y)] = E(X)$$

A proof of Proposition 5.3.1 will be given in Section 5.4, Corollary 1. The following example may make the proposition appear more plausible.

Example 3 Let X and Y have a discrete joint distribution $P_{X,Y}$ and marginal distributions P_X and P_Y. We know that $E(Y) = \sum_y y P_Y(y)$ where y ranges the discrete points of P_Y. Now by the partition formula (Proposition 2.2.2) we have

$$P_Y(y) = \sum_x P_X(x) P_{Y|x}(y)$$

where x ranges the discrete points of P_X. Hence,

$$E(Y) = \sum_y y P_Y(y)$$

$$= \sum_y y \left[\sum_x P_X(x) P_{Y|x}(y) \right]$$

$$= \sum_x P_X(x) \left[\sum_y y P_{Y|x}(y) \right]$$

$$= \sum_x P_X(x)[E(Y|x)]$$

$$= E[E(Y|X)]$$

Proposition 5.3.1 has the following useful corollary, which we need to prove Theorem 5.3.1. Imagine a Borel function g of two variables. If X and Y are jointly distributed, then so are X and $g(X, Y)$. Consequently, just as it made sense to talk about the random variable $E(Y|X)$, it now makes sense to talk about the random variable $E[g(X, Y)|X]$ as well as its expected value $E[E(g(X, Y)|X)]$, and on account of Proposition 5.3.1 we may state the following corollary.

5.3.1 COROLLARY

Given jointly distributed random variables X and Y, and a Borel function g of two variables, we have

$$E[E(g(X, Y)|X)] = E[g(X, Y)]$$

and likewise

$$E[E(g(X, Y)|Y)] = E[g(X, Y)]$$

Returning to the question of relation between the regression coefficients and correlation coefficient, we now state the following theorem, which says essentially that if Y has a linear regression on X then the regression coefficients can be determined by various numerical characterizations of the random vector (X, Y) considered in the preceding section including the correlation coefficient.

5.3.1 THEOREM

Given jointly distributed random variables X and Y, if Y has a linear regression on X, say, $E(Y|x) = \alpha + \beta x$, then

$$\alpha = \mu_Y - \beta \mu_X$$

$$\beta = \frac{\sigma_Y}{\sigma_X} \rho$$

where μ_X, μ_Y, $\sigma_X > 0$, $\sigma_Y > 0$, and ρ are the mean values, the standard deviations, and correlation coefficient of X and Y, respectively.

PROOF

$E(Y|x) = \alpha + \beta x$ leads to the random variable

$$E(Y|X) = \alpha + \beta X$$

By Proposition 5.3.1

$$E[E(Y|X)] = E(Y) = \mu_Y$$

On the other hand

$$E(\alpha + \beta X) = \alpha + \beta \mu_X$$

Consequently we have

$$\alpha + \beta \mu_X = \mu_Y \tag{1}$$

Also $E(xY|x) = \alpha x + \beta x^2$ leads to the random variable

$$E(XY|X) = \alpha X + \beta X^2$$

Now by Corollary 5.3.1

$$E[E(XY|X)] = E(XY) = \text{cov}(X, Y) - \mu_X \mu_Y$$

On the other hand

$$E(\alpha X + \beta X^2) = \alpha \mu_X + \beta(\sigma_X^2 + \mu_X^2)$$

Consequently we have

$$\alpha \mu_X + \beta(\sigma_X^2 + \mu_X^2) = \sigma_X \sigma_Y \rho - \mu_X \mu_Y \tag{2}$$

Solving (1) and (2) for β and α, we obtain

$$\beta = \frac{\sigma_Y}{\sigma_X} \rho$$

and

$$\alpha = \mu_Y - \beta \mu_X$$

$$= \mu_Y - \frac{\sigma_Y}{\sigma_X} \rho \mu_X$$

This completes the proof.

If we substitute these values for α and β in the regression equation, $y = \alpha + \beta x$, we obtain

$$y = \mu_Y + \frac{\sigma_Y}{\sigma_X} \rho(x - \mu_X)$$

or

$$(y - \mu_Y) = \frac{\sigma_Y}{\sigma_X} \rho(x - \mu_X)$$

In other words we have the following corollary.

5.3.2 COROLLARY

Given jointly distributed random variables X and Y having all the usual numerical characterizations, μ_X, μ_Y, σ_X, σ_Y, and ρ, if Y has a linear regression on X, then the regression line will pass the point (μ_X, μ_Y) and have a slope equal to $(\sigma_Y/\sigma_X)\rho$.

If Y has a linear regression on X, this does not necessarily mean X will have a linear regression on Y (see Problem 5.3.2). However, if X does have a linear regression on Y, then it is clear that the regression equation will be $(x - \mu_X) = (\sigma_X/\sigma_Y)\rho(y - \mu_Y)$.

We digress slightly to consider the following problem:

Equation (1) in the proof of Theorem 5.3.1

$$\mu_Y = \alpha + \beta\mu_X$$

is significant in that it says that if Y has a linear regression on X then μ_Y can be calculated from μ_X via regression coefficients. This leads us to wonder whether it is likewise possible to calculate σ_Y from σ_X via regression coefficients if Y has a linear regression on X. The answer is a conditional yes, as shown in Theorem 5.3.2. This theorem makes sense provided that we know what is meant by *conditional variance*.

5.3.3 DEFINITION

For jointly distributed random variables X and Y the *conditional mean* and the *conditional variance* of Y given $X = x$ are defined respectively by

$$\mu_{Y|x} = E(Y|x)$$
$$\sigma^2_{Y|x} = E[(Y - \mu_{Y|x})^2|x]$$

The precise meaning of $E[(Y - \mu_{Y|x})^2|x]$ is simply the conditional expected value $E(Z|x)$ where X and $Z = (Y - \mu_{Y|x})^2$ are two jointly distributed random variables.

In Example 2 of this section, both the conditional mean and the conditional variance can be read off the conditional density function, which is $(\rho x, \sqrt{1 - \rho^2})$ normal. Thus,

$$\mu_{Y|x} = \rho x$$
$$\sigma^2_{Y|x} = 1 - \rho^2$$

5.3.2 THEOREM

Given jointly distributed random variables X and Y, if Y has a linear regression on X, say, $E(Y|x) = \alpha + \beta x$, and furthermore if the conditional variance remains unchanged, say, $\mathrm{var}(Y|x) = \sigma^2$ for all x where σ^2 is a constant, then

$$\sigma^2_Y = \beta^2\sigma^2_X + \sigma^2$$

PROOF

The equation $\text{var}(Y|x) = E[(Y - \alpha - \beta x)^2|x]$ leads to the random variable

$$E[(Y - \alpha - \beta X)^2|X] = \text{var}(Y|X) \tag{1}$$

which is really only a constant since $\text{var}(Y|x) = \sigma^2$ for all x. Thus $E[\text{var}(Y|X)] = \sigma^2$. On the other hand, by Corollary 5.3.1 we have the expected value of the left-hand side of (1) equal to

$$E[(Y - \alpha - \beta X)^2] = E(Y^2 + \alpha^2 + \beta^2 X^2 - 2\alpha Y - 2\alpha\beta X - 2\beta XY)$$

$$= (\sigma_Y^2 + \mu_Y^2) + \alpha^2 + \beta^2(\sigma_X^2 + \mu_X^2) - 2\alpha\mu_Y - 2\alpha\beta\mu_X - 2\beta(\sigma_X\sigma_Y\rho + \mu_X\mu_Y)$$

$$= \sigma_Y^2 + \beta^2\sigma_X^2 - 2\beta\sigma_X\sigma_Y\rho + (\mu_Y - \alpha - \beta\mu_X)^2$$

$$= \sigma_Y^2 + \beta^2\sigma_X^2 - 2\beta\sigma_X\sigma_Y\left(\beta\frac{\sigma_X}{\sigma_Y}\right)$$

since $\mu_Y = \alpha + \beta\mu_X$ and $\rho = \beta(\sigma_X/\sigma_Y)$ by Theorem 5.3.1. After simplification this last expression reduces to $\sigma_Y^2 - \beta_X^2\sigma_X$, which is supposed to be equal to σ^2. Consequently we obtain

$$\sigma_Y^2 = \beta^2\sigma_X^2 + \sigma^2$$

This completes the proof.

Note that if $\sigma^2 = 0$ in the above equation, we have $\sigma_Y^2 = \beta\sigma_X^2$. This stands to reason since if $\sigma^2 = \text{var}(Y|x) = 0$ this means Y is a constant given $X = x$ for all x, so that $Y = \alpha + \beta X$ exactly, and of course

$$\text{var}(Y) = \text{var}(\alpha + \beta X) = \beta^2\sigma_X^2$$

We conclude this section with a consideration regarding applications of regression coefficients in prediction. We recall that regression coefficients were originally introduced only for those pairs of random variables in which one has a linear regression on the other. Later, however, we have succeeded in relating regression coefficients to correlation coefficients, which, unlike regression coefficients, are defined for pairs of random variables including those not having any linear regression. This means we may artificially define, if we wish, regression coefficients for an arbitrary (ordered) pair of random variables X and Y by setting

$$\alpha = \mu_Y - \beta\mu_X$$

$$\beta = \frac{\sigma_Y}{\sigma_X}\rho$$

The question is whether this is worth doing. The answer is yes, and we will explain why.

Given jointly distributed X and Y, if Y is known to have a linear regression on X, that is:

$$E(\bar{Y}|x) = \alpha + \beta x$$

then although corresponding to each value $X = x$ many values of Y are in general possible, in the absence of any information regarding these values of Y the most sensible thing to do is to guess (or predict) that the corresponding value of Y is $\alpha + \beta x$. If we let this predicted value of Y be $\hat{Y}$, then $\hat{Y}$ is a random variable which is in fact a function of the random variable X, that is:

$$\hat{Y} = \alpha + \beta X$$

If, Y does not have a linear regression on X and somehow it is desirable to have a simple scheme of predicting the value of $Y(\omega)$ corresponding to the observed value $X(\omega) = x$, we may set

$$\hat{Y} = a + bX$$

and proceed to adjust the constants a and b so as to make the predicted value $\hat{Y}$ least erroneous on the average. That is, we want a and b to be so chosen that the expected value

$$E[(Y - \hat{Y})^2] = \delta$$

will be smallest possible. Now this can be accomplished by expressing the expected value δ explicitly in terms of a and b and then deciding what a and b should be in order to minimize δ. Accordingly we proceed as follows (this method is known as the method of least squares).

$$
\begin{aligned}
\delta &= E(Y - \hat{Y})^2 = E(Y - a - bX)^2 \\
&= E[(Y - \mu_Y) - b(X - \mu_X) + (\mu_Y - a - b\mu_X)^2] \\
&= E(Y - \mu_Y)^2 + b^2 E(X - \mu_X)^2 + (\mu_Y - a - b\mu_X)^2 - 2bE[(X - \mu_X)(Y - \mu_Y)] \\
&= \text{var}(Y) + b^2 \, \text{var}(X) + (\mu_Y - a - b\mu_X)^2 - 2b \, \text{cov}(X, Y) \\
&= \sigma_Y^2 + b^2\sigma_X^2 - 2b\sigma_X\sigma_Y\rho + (\mu_Y - a - b\mu_X)^2 \\
&= (b\sigma_X - \rho\sigma_Y)^2 + \sigma_Y^2(1 - \rho^2) + (\mu_Y - a - b\mu_X)^2
\end{aligned}
$$

Hence it follows that δ is at its minimum when

$$b\sigma_X - \rho\sigma_Y = 0$$
$$\mu_Y - a - b\mu_X = 0$$

Solving these equations we obtain

$$b = \frac{\sigma_Y}{\sigma_X}\rho$$

$$a = \mu_Y - \beta\mu_X$$

The solutions of a and b look familiar since they are exactly the regression coefficients artificially defined a while ago. In other words for "linear prediction" of Y from X we are better off in the long run (or on the average) to set

$$\hat{Y} = \alpha + \beta X$$

where $\beta = (\sigma_X/\sigma_Y)\rho$ and $\alpha = \mu_Y - \beta\mu_X$ even though Y does not have a linear regression on X.

We summarize the preceding discussion as follows.

5.3.4 DEFINITION

Given a pair of jointly distributed random variables X and Y, we define the regression coefficients of Y on X by

$$\alpha = \mu_Y - \beta\mu_X$$

$$\beta = \frac{\sigma_X}{\sigma_Y}\rho$$

where μ_X, μ_Y, σ_X, σ_Y, and ρ are the mean values of X and Y, the standard deviations of X and Y, and the correlation coefficients of X and Y, respectively.

5.3.3 THEOREM

Given a pair of jointly distributed random variables X and Y, if we predict the value of Y on the basis of the value of X by $\hat{Y} = a + bX$, then the mean error $E(Y - \hat{Y})^2$ is at its minimum when $a = \alpha$ and $b = \beta$, where α and β are regression coefficients defined in Definition 5.3.4.

5.3.5 DEFINITION

Given a pair of jointly distributed random variables X and Y, the random variable $\hat{Y}$ given by $\hat{Y} = \alpha + \beta X$, where $\alpha = \mu_Y - \beta\mu_X$ and $\beta = (\sigma_X/\sigma_Y)\rho$ are the regression coefficients of Y on X, is called the *best linear prediction* of Y by X.

5.3 PROBLEMS

1. If X and Y are jointly distributed over the half disk of radius 1 above the x-axis centered at the origin, sketch the graphs of $y = E(Y|x)$ and $x = E(X|y)$.

2. If X and Y are jointly distributed over the unit square with the density function $f(x, y) = \frac{2}{3}(x + 2y)$ for $0 \leq x \leq 1$ and $0 \leq y \leq 1$, calculate $E(Y|x)$ and $\text{var}(Y|x)$.

3. A thick coin falls heads with probability $p > 0$, falls tails with probability $q > 0$, and falls "sides" with probability $1 - p - q > 0$. Let X, Y, and W be respectively the random variables counting the number of heads, tails, and sides in n tosses of the coin. Plot the joint distribution of X and W, and calculate $E(X|W = k)$ and $\text{var}(X|W = k)$ for $0 \leq k \leq n$.

4. Let X and Y respectively represent the height (in feet) and the weight (in pounds) of female adults in a certain population. Further suppose the following numerical characteristics about X and Y are known: $\mu_X = 5$ ft, $\sigma_X = 0.25$ ft, $\mu_Y = 130$ lb, $\sigma_Y = 10$ lb, $E(XY) = 651.25$ ft lb

    a. Calculate the covariance and the correlation coefficient (ρ) of X and Y.

    b. Determine the regression coefficients (α and β) of Y on X.

    c. If a certain female adult chosen at random from the population has height 5.5 ft, what is the best linear prediction ($\hat{Y}$) of her weight based on her height (X)?

5. Let (X, Y) be such that X and Y are independent and each has a standard normal distribution. Let (X, Y) be transformed into (V, W) by

$$V = X$$
$$W = \beta X + Y \qquad 0 < \beta < 1$$

Determine the mean values, standard deviations, and the correlation coefficient of V and W.

6. Let (X, Y) be as given in Problem 5.3.5. If (X, Y) is transformed into (V, W) by

$$V = X$$
$$W = \beta X + \gamma Y \qquad \text{with} \quad \beta = \tfrac{1}{2}$$

determine γ so that $\sigma_W = 1$. Then calculate the correlation coefficient ρ of V and W.

7. Let (X, Y) be as given in Problem 5.3.5 again. If (X, Y) is transformed into (V, W) by

$$V = X$$
$$W = \rho X + \sqrt{1 - \rho^2}\, Y$$

show that the density function of (V, W) is

$$f_{V,\,W}(x, y) = \frac{1}{2\pi\sqrt{1 - \rho^2}} \exp\left[-\frac{1}{2(1 - \rho^2)} (x^2 - 2\rho xy + y^2) \right]$$

5.4 ABSTRACT CONDITIONAL EXPECTATIONS

Given a random variable X over a probability space $(S, \mathscr{S}, P)$ the mathematical expectation of X was defined as the constant μ such that

$$\mu = \int_S X(\omega)\, P(d\omega) \tag{1}$$

Now if we regard μ as a constant random variable

$$\mu(\omega) = \mu \qquad \text{for all} \quad \omega \text{ in } S$$

then we may rewrite (1) as

$$\int_S \mu(\omega)\, P(d\omega) = \int_S X(\omega)\, P(d\omega)$$

The constant random variable $\mu(\omega)$ then can be regarded as an averaging (or leveling) of $X(\omega)$ over the entire S. This observation suggests the possibility of averaging $X(\omega)$ not all at once over the entire S, but rather piecewise over the subsets (events) $A_1, A_2, \ldots, A_n$ forming a partition of S. Specifically consider the random variable $v(\omega)$ over $(S, \mathscr{S}, P)$ such that

 (i) $v(\omega) = a_i$ on each A_i where a_i is a constant
(ii) $\int_{A_i} v(\omega)\, P(d\omega) = \int_{A_i} X(\omega)\, P(d\omega)$ for each A_i

Clearly $v(\omega)$ can be regarded as a piecewise averaging of $X(\omega)$ relative to $\{A_1, A_2, \ldots, A_n\}$. Now if we let $\mathscr{A}$ be the smallest sigma (or Boolean in this case) algebra of subsets of S containing all the $A_1, A_2, \ldots, A_n$ (so that clearly $\mathscr{A} \subset \mathscr{S}$), then $v(\omega)$ can be characterized as a random variable over the probability space $(S, \mathscr{A}, P)$ such that

$$\int_A v(\omega)\, P(d\omega) = \int_A X(\omega)\, P(d\omega)$$

for any member A of $\mathscr{A}$. Thus we may view $v(\omega)$ as the conditional expectation of $X(\omega)$ given the sigma algebra $\mathscr{A}$. This brings us to the following definition.

5.4.1 DEFINITION
Given a random variable X over the probability space $(S, \mathscr{S}, P)$, by the *conditional expectation of X given a sigma algebra $\mathscr{A} \subset \mathscr{S}$*, we mean the random variable $\mu_{X|\mathscr{A}}$ over the probability space $(S, \mathscr{A}, P)$ satisfying

$$\int_A \mu_{X|\mathscr{A}}(\omega)\, P(d\omega) = \int_A X(\omega)\, P(d\omega)$$

for each A in $\mathscr{A}$. We shall also use the notation $\mu_{\mathscr{A}}$ for $\mu_{X|\mathscr{A}}$ when there is no fear of confusion.

Note that if $\mathscr{A}$ happens to be the trivial sigma algebra consisting of two sets ϕ and S, then the conditional expectation $\mu_{\mathscr{A}}$ reduces to the ordinary expectation μ. The existence of $\mu_{\mathscr{A}}$ as described in the above definition can be proved by using the Radon-Nikodým theorem (Theorem 3.5.1). Further details may be found in Loève's book (see Minibibliography).

The preceding definition of conditional expectation of X given $\mathscr{A}$ allows us to consider as an important special case the conditional expectation of X given Y where X and Y are (jointly distributed) random variables defined over the same probability space. The point is, Y gives rise to the following sigma algebra $\mathscr{A}$ relative to which the conditional expectation of X may be considered:

$$\mathscr{A} = \{Y^{-1}(B): B \in \mathscr{B}\}$$

where $\mathscr{B}$ is the Borel class of sets in the range R_Y of Y. For later convenience in notation, however, we shall give the definition of the conditional expectation of Y given X instead of that of X given Y.

5.4.2 DEFINITION

Given random variables X and Y on a probability space $(S, \mathscr{S}, P)$ by the *conditional expectation of Y given X*, denoted by $\mu_{Y|X}$, we mean the conditional expectation of Y given $\mathscr{A}$ where

$$\mathscr{A} = \{X^{-1}(B): B \in \mathscr{B}\}$$

Now it is not difficult to see that $\mu_{Y|X}(\omega)$ has to remain constant over each set A of the form $A = X^{-1}(x)$ where x is a single point in the range of X, for otherwise $\mu_{Y|X}$ will not quite make a random variable over $(S, \mathscr{A}, P)$. We shall denote this constant by $E(Y|x)$, that is:

$$E(Y|x) = \mu_{Y|X}(\omega) \qquad \text{for any} \quad \omega \text{ in } X^{-1}(x)$$

Since the constant $E(Y|x)$ is clearly dependent on x, we have thus a function $g(x) = E(Y|x)$, and the function $y = g(x)$ may be graphed in the $R_X \times R_Y$ plane, giving what is known as the regression graph (or curve) of Y on X (see Fig. 5.4.1). We summarize this as follows.

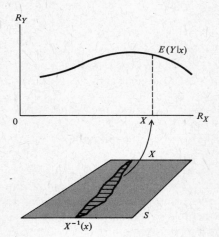

Figure 5.4.1

5.4.3 DEFINITION

Given jointly distributed random variables X and Y, that is, X and Y are both defined on the same probability space $(S, \mathscr{S}, P)$, by the regression *graph* of Y on X we mean the following subset of $R_X \times R_Y$

$$\{(x, E(Y|x)): x \in X(S) \subset R_X\}$$

where $E(Y|x) = \mu_{Y|X}(\omega)$ with $\omega \in X^{-1}(x)$ is called the *conditional expected value of Y given X = x.*

We now begin the discussion that will lead to a simple proof of the relation $E[E(Y|X)] = E(Y)$ given in Proposition 5.3.1.

Given X on $(S, \mathcal{S}, P)$, the conditional expectation of X given $\mathcal{A} \subset \mathcal{S}$ will look different if different sigma algebras $\mathcal{A}$ are considered. In particular, if $\mathcal{A}_0 = \{\phi, S\}$, then as pointed out earlier $\mu_{\mathcal{A}_0}$ is a mere constant giving the ordinary expected value μ of X. On the other hand if $\mathcal{A}_1 = \{X^{-1}(B): B \in \mathcal{B}\}$ where $\mathcal{B}$ is the Borel class of subsets in the range of X, then $\mu_{\mathcal{A}_1} = X$ since X itself is clearly a random variable over $(S, \mathcal{A}_1, P)$ and the equation $\int_A \mu_{\mathcal{A}_1}(\omega)\, P(d\omega) = \int_A X(\omega)\, P(d\omega)$ for every $A \in \mathcal{A}_1$ is trivially satisfied by letting $\mu_{\mathcal{A}_1} = X$. In between these two extreme cases we now consider $\mu_{\mathcal{A}}$ where $\mathcal{A}$ is a sigma algebra such that $\mathcal{A}_0 \subset \mathcal{A} \subset \mathcal{A}_1$.

The conditional expectation $\mu_{\mathcal{A}}$ is a random variable on $(S, \mathcal{A}, P)$, but since $\mathcal{A} \subset \mathcal{S}$ this makes $\mu_{\mathcal{A}}$ easily a random variable on $(S, \mathcal{S}, P)$, and we may consider the ordinary mathematical expectation of $\mu_{\mathcal{A}}$ as a random variable on $(S, \mathcal{S}, P)$. That is, after having piecewise averaged X we now want to average it completely after all. We should then expect to have the ordinary expected value of X in the end. The following proposition says precisely this.

5.4.1 PROPOSITION

Given a random variable X on $(S, \mathcal{S}, P)$ and a sigma algebra $\mathcal{A} \subset \mathcal{S}$, we have

$$E(\mu_{\mathcal{A}}) = E(X)$$

PROOF

We have

$$E(\mu_{\mathcal{A}}) = \int_S \mu_{\mathcal{A}}(\omega)\, P(d\omega)$$

but

$$\int_A \mu_{\mathcal{A}}(\omega)\, P(d\omega) = \int_A X(\omega)\, P(d\omega)$$

for any $A \in \mathcal{A}$ by Definition 5.4.1. In particular, taking $A = S$, we obtain

$$\int_S \mu_{\mathcal{A}}(\omega)\, P(d\omega) = E(X)$$

completing the proof.

As an important special case of the proposition we have the following corollary.

5.4.1 COROLLARY

Given X and Y on $(S, \mathscr{S}, P)$ we have

$$E(\mu_{Y|X}) = E(Y)$$

or alternatively,

$$E[E(Y|X)] = E(Y)$$

Corollary 5.4.1 implies the following corollary.

5.4.2 COROLLARY

If Y has a partly discrete and partly absolutely continuous distribution, say

$$\sum_i P_Y(y_i) + \int_{-\infty}^{\infty} f(y)\, dy = 1$$

where $y_1, y_2, \ldots$ are the discrete points and $f(y)$ is the (partial) density function, then

$$E(Y) = \sum_i y_i P_Y(y_i) + \int_{-\infty}^{\infty} y f(y)\, dy$$

Although this last formula can be seen to be a direct consequence of the general definition of mathematical expectation (cf. Section 4.1), it may also be regarded as a consequence of the relation $E(Y) = E[E(Y|X)]$, as follows.

PROOF

Let X be a random variable that takes the values 0 and 1 with the probabilities p and q, respectively, where

$$p = \sum_i P_Y(y_i)$$

$$q = \int_{-\infty}^{\infty} f(y)\, dy$$

Consider the conditional probabilities:

$$P_{Y|x=0}(y_i) = \frac{P_Y(y_i)}{p}$$

$$P_{Y|x=1}(B) = \int_B \frac{f(y)}{q}\, dy$$

That is, $P_{Y|x=0}$ is discrete with the discrete points $y_1, y_2, \ldots$ and $P_{Y|x=1}$ has a density function $f(y)/q$. Then the joint distribution of X and Y is such that the marginal distribution of Y is exactly as given in the corollary. Now, the random variable $g(X) = E(Y|X)$ has a two-point distribution at $g(0) = E(Y|x = 0)$ and $g(1) = E(Y|x = 1)$ with probabilities p and q, hence by Corollary 5.4.1

$$E(Y) = E[E(Y|X)]$$
$$= pE(Y|x = 0) + qE(Y|x = 1)$$

But

$$E(Y|x = 0) = \sum_i y_i \frac{P_Y(y_i)}{p}$$

and

$$E(Y|x = 1) = \int_{-\infty}^{\infty} y \frac{f(y)}{q} \, dy$$

Consequently,

$$E(Y) = \sum_i y_i P_Y(y_i) + \int_{-\infty}^{\infty} y f(y) \, dy$$

Example 1 Let Y be discretely distributed at -1 and 2, with probabilities $\frac{1}{2}$ and $\frac{1}{4}$, respectively, and be continuously distributed over the interval $[-1, 3]$ with a uniform density $f(y) = \frac{1}{16}$. Calculate $E(Y)$ and $E(Y^3)$.

Solution By Corollary 5.4.2 we have

$$E(Y) = (-1) \cdot \tfrac{1}{2} + 2 \cdot \tfrac{1}{4} + \int_{-1}^{3} y \cdot \tfrac{1}{16} \, dy = \tfrac{1}{4}$$

$$E(Y^3) = (-1)^3 \cdot \tfrac{1}{2} + 2^3 \cdot \tfrac{1}{4} + \int_{-1}^{3} y^3 \cdot \tfrac{1}{16} \, dy = \tfrac{11}{14}$$

5.4 PROBLEMS

1. Let X and Y be random variables defined on a probability space $(S, \mathscr{S}, P)$ such that their joint distribution has exactly three discrete points at $(1, 1)$, $(1, 2)$, and $(2, 3)$, each with probability $\frac{1}{3}$. Let $\mathscr{A} \subset \mathscr{S}$ be the smallest sigma algebra containing the sets $X^{-1}(1)$ and $X^{-1}(2)$. Show that Y is not a random variable over the probability space $(\mathscr{S}, \mathscr{A}, P)$. If however a new random variable $\hat{Y}$ is defined by

$$\hat{Y}(\omega) = \begin{cases} \tfrac{3}{2} & \text{whenever} \quad X(\omega) = 1 \\ 3 & \text{whenever} \quad X(\omega) = 2 \end{cases}$$

then show that $\hat{Y}$ is a random variable over $(S, \mathscr{A}, P)$.

2. Given a set S together with a sigma algebra $\mathscr{A}$ of its subsets, a subset $A \subset S$ is said to be a $\mathscr{A}$-atom if A is a member of $\mathscr{A}$ but any proper subset of A is not. Show that if X is a random variable on $(S, \mathscr{S}, P)$ and $A \subset \mathscr{S}$ is a sigma algebra, then the conditional expectation $X|_{\mathscr{A}}$ assumes a constant value over any $\mathscr{A}$-atom, that is:

$$X|_{\mathscr{A}}(\omega_1) = X|_{\mathscr{A}}(\omega_2)$$

whenever ω_1 and ω_2 are contained in the same $\mathscr{A}$-atom.

3. Let X be a random variable on $(S, \mathscr{S}, P)$ and $\hat{X}$ the conditional expectation of X given the sigma algebra $\mathscr{A} \subset \mathscr{S}$. Show that the value of $\hat{X}$ over any $\mathscr{A}$-atom of probability 0 may be altered to any other constant without affecting the fact that $\hat{X}$ is a conditional expectation of X given $\mathscr{A}$.

4. Let X be a random variable on $(S, \mathscr{S}, P)$ and $\mathscr{A}_1 \subset \mathscr{A}_2 \subset \mathscr{S}$ be two subsigma algebras. If $\mu_{X|\mathscr{A}_1}$ and $\mu_{X|\mathscr{A}_2}$ are respectively the conditional expectations of X given $\mathscr{A}_1$ and $\mathscr{A}_2$, show that the conditional expectation of $\mu_{X|\mathscr{A}_2}$ given $\mathscr{A}_1$ is exactly $\mu_{X|\mathscr{A}_1}$.

Chapter 6
LAWS OF LARGE
NUMBERS AND
LIMIT
THEOREMS

This chapter is concerned with essentially two types of limit theorems although the first type is known under the mathematically unusual name of *laws* of large numbers. A more descriptive (but less convenient) name might be the "limit theorem for the averages of random variables." The second type is known as central limit theorems though again a more descriptive name might be the "limit theorem for the distributions of standardized sums of random variables."

Section 6.1 (Chebyshev Inequality and the Weak Law of Large Numbers) and 6.2 (The Strong Law of Large Numbers) treat the so-called *laws*. Actually, what we have are two *mathematical* theorems formalizing some alleged *physical* laws. Section 6.5 (Central Limit Theorems) contains a family of theorems regarded to be of central importance by probabilists and statisticians. The theory of statistics often begins with a central limit theorem, using it as the main justification for assuming just about any random variable arising from sampling as normally distributed. Thus, there is no reason why one should not refer to central limit theorems as laws of some sort, though this is not generally done.

Sections 6.3 (Characteristic Functions) and 6.4 (Convergence in Distribution) contain concepts and tools preparatory to Section 6.5. Some propositions and theorems in these two sections are not proved; however, they are made comprehensible with lead-in material.

6.1 CHEBYSHEV INEQUALITY AND THE WEAK LAW OF LARGE NUMBERS

Mathematical models for a coin can be a circle, a cylinder, or a random variable with a two-point distribution, depending on whether we are interested in its shape, volume, or falling heads and tails. As long as we are only interested in its falling heads or tails, we shall let a random variable X such that $P_X(1) = p$ for some $0 < p < 1$ and $P_X(0) = 1 - p$ be the basic mathematical model for the given coin. If this coin is tossed a large number n of times, the appropriate mathematical model for the total number of heads in these n tosses is the random variable

$$S_n = X_1 + X_2 + \cdots + X_n$$

where the X_i are independent and have distributions identical to that of X. Furthermore, if we are interested in the average number of heads (i.e., the ratio of heads to tosses) in these n tosses, the appropriate mathematical model should be the random variable

$$A_n = \frac{S_n}{n} = \frac{1}{n}(X_1 + X_2 + \cdots + X_n)$$

Now since we generally suspect (and experience bears this out) that the average number of heads in a series of n tosses shouldn't vary too much from one series of n tosses to another provided that n is a fixed large number, we want to investigate if the corresponding mathematical model A_n will bear this out also. The result of this investigation is one form of the so-called weak law of large numbers, the principal tool of investigation being the celebrated Chebyshev inequality.

We now proceed somewhat more formally as follows.

6.1.1 DEFINITION

An n-tuple of independent random variables $(X_1, X_2, \ldots, X_n)$ each having an identical distribution as some fixed random variable X is called the *sample* of size n for X. The random variable

$$A_n = \frac{1}{n}(X_1 + X_2 + \cdots + X_n)$$

is then called the *sample average* of size n of X. Parenthetically, we might call the random variable

$$S_n = X_1 + X_2 + \cdots + X_n$$

the *sample sum* of size n of X.

The sample average A_n, which is a random variable in its own right, has the same mean value as X and a much smaller standard deviation than X, especially when n is a larger number. We will see this in the following proposition.

6.1.1 PROPOSITION

If X has a finite mean value μ and a finite standard deviation σ, then the sample average A_n of X also has a finite mean and a finite standard deviation, in fact

$$\mu_{A_n} = \mu$$

$$\sigma_{A_n} = \frac{\sigma}{\sqrt{n}}$$

In particular, if $P_X(1) = p$ and $P_X(0) = q$ so that $\mu = p$ and $\sigma = \sqrt{pq}$, then $\mu_{A_n} = p$ and $\sigma_{A_n} = \sqrt{pq/n}$.

PROOF
We need only calculate μ_{A_n} and σ_{A_n} by making use of the basic properties of mean values (μ) and variances (σ^2) listed in Section 4.3.

$$\mu_{A_n} = \mu_{(1/n)(X_1 + \cdots + X_n)} = \frac{1}{n}(\mu_{X_1} + \cdots + \mu_{X_n}) = \frac{1}{n}(n\mu) = \mu$$

$$\sigma_{A_n}^2 = \sigma_{(1/n)(X_1 + \cdots + X_n)}^2 = \frac{1}{n^2}(\sigma_{X_1}^2 + \cdots + \sigma_{X_n}^2) = \frac{1}{n^2}(n\sigma^2) = \frac{\sigma^2}{n}$$

Consequently,

$$\sigma_{A_n} = \frac{\sigma}{\sqrt{n}}$$

If a random variable has a small standard deviation σ, its probability mass tends to cluster around the mean value μ. Chebyshev's inequality says that if you take an interval of radius $n\sigma$ around the mean value μ, $(\mu - n\sigma, \mu + n\sigma)$, you will find outside of this interval at most $1/n^2$ amount of probability mass; thus, for example, for $n = 2$ we have

$$P(|X - \mu| \geqq 2\sigma) \leqq \tfrac{1}{4}$$

6.1.2 PROPOSITION (Chebyshev Inequality)
If μ and σ are the mean value and the standard deviation of X, then for any positive real number n we have

$$P(|X - \mu| \geqq n\sigma) \leqq \frac{1}{n^2}$$

PROOF
We recall that

$$\sigma^2 = E[(X - \mu)^2]$$

In examining the expected value for the random variable

$$Y = (X - \mu)^2$$

we note that Y is always nonnegative and takes values at least $(n\sigma)^2$ whenever X is at least $n\sigma$ distance away from μ. Now compare Y with another random variable Z which takes exactly the value $(n\sigma)^2$ whenever X is at least $n\sigma$ away from μ and takes the value 0 otherwise. Clearly

$$Z \leqq Y$$

Hence

$$E(Z) \leq E[(X - \mu)^2] = \sigma^2$$

but

$$E(Z) = (n\sigma)^2 \cdot P(|X - \mu| \geq n\sigma) + 0 \cdot P(|X - \mu| < n\sigma)$$

from which it follows that

$$P(|X - \mu| \geq n\sigma) = E(Z)/(n\sigma^2) \leq \frac{1}{n^2}$$

In view of the Chebyshev inequality we see that since σ_{A_n} approaches 0 as n approaches ∞, for a very large n the probability mass of A_n must be pretty heavily concentrated around its mean value. Formally, we state the following theorem.

6.1.1 THEOREM (A Weak Law of Large Numbers)

If A_n is the sample average of X, and the mean value μ and the standard deviation σ of X are both finite, then

$$\lim_{n \to \infty} P(|A_n - \mu| \geq \varepsilon) = 0$$

for any $\varepsilon > 0$ however small.

PROOF

According to Chebyshev's inequality (Proposition 6.1.2) we have

$$P(|A_n - \mu| \geq \varepsilon) = P(|A_n - \mu| \geq (\varepsilon/\sigma_{A_n}) \cdot \sigma_{A_n})$$

$$\leq \frac{\sigma_{A_n}^2}{\varepsilon^2}$$

but $\sigma_{A_n}^2 \to 0$ since $\sigma_{A_n} = \sigma/\sqrt{n}$ by Proposition 6.1.1. Hence $P(|A_n - \mu| \geq \varepsilon) \to 0$ and the theorem is proved.

A special case of this theorem in which X has a two-point distribution $P_X(1) = p$ and $P_X(0) = q$ was originally found by Jacob Bernoulli (1713).

Note that $P(|A_n - \mu| \geq \varepsilon)$ approaching 0 is equivalent to $P(|A_n - \mu| < \varepsilon)$ approaching 1. The interpretation of Theorem 6.1.1 in terms of coin tossing (i.e., X has the two-point distribution) is as follows: If a coin is tossed a large number n of times, the average number of heads A_n (i.e., the ratio of heads to n) will stay close to some fixed number $\mu = p$ (but we don't know how big this number is exactly) in the sense that A_n will be within ε distance from this fixed number with probability nearly 1, say, $1 - \delta$. As to how nearly 1 this probability is for some given ε, this will depend on how large an n we are willing to take. In general, for a prescribed small ε and δ we can calculate how large n has to be.

Theorem 6.1.1. suggests the following abstract definition which leads to possible generalizations of the theorem.

6.1.2 DEFINITION
A sequence of random variables $X_1, X_2, X_3, \ldots$, is said to obey the weak law of large numbers if

$$\lim_{n \to \infty} P(|A_n - a_n| < \varepsilon) = 1$$

for any $\varepsilon > 0$ where

$$A_n = \frac{1}{n}(X_1 + X_2 + \cdots + X_n)$$

$$a_n = \frac{1}{n}(\mu_{X_1} + \mu_{X_2} + \cdots + \mu_{X_n})$$

In words, the "random average" A_n approaches the "means average" a_n "in probability" as n approaches ∞, or $A_n - a_n$ approaches 0 in probability as n approaches ∞. According to this definition, Theorem 6.1.1 simply says that $X_1, X_2, X_3, \ldots$ obeys the weak law of large numbers if
(1) $X_1, X_2, X_3, \ldots$ are independent.
(2) $X_1, X_2, X_3, \ldots$ are identically distributed as some X.
(3) X and hence each X_i has a finite mean and a finite variance.

Thus, Theorem 6.1.1 says that (1), (2), (3) above constitute a sufficient set of conditions for $X_1, X_2, X_3, \ldots$ to obey the weak law of large numbers. However, it turns out that hardly any of these conditions is necessary, and this leads to various generalizations of Theorem 6.1.1. In fact we have the following four well-known results.

6.1.2 THEOREM (Chebyshev)
The sequence $X_1, X_2, X_3, \ldots$ obeys the weak law of large numbers if
(1) $X_1, X_2, X_3, \ldots$ are independent.
(2) $\sigma_{x_i}^2 \leq C$ for all i for some fixed bound C.

PROOF
As in the proof of Theorem 6.1.1 we have in view of the Chebyshev inequality

$$P(|A_n - a_n| \geq \varepsilon) \leq \frac{\sigma_{A_n}^2}{\varepsilon^2}$$

but

$$\sigma_{A_n}^2 = \frac{1}{n^2} \sum_{i=1}^{n} \mathrm{var}(X_i) \leq \frac{1}{n^2}(nC) = C/n \to 0.$$

Hence $P(A_n - a_n| \geq \varepsilon) \to 0$ and the theorem is proved.

The proofs of Theorems 6.1.1 and 6.1.2 clearly indicate that the essential condition for $X_1, X_2, X_3, \ldots$ to obey the weak law of large numbers is that $\sigma_{A_n}^2 \to 0$ since all the given conditions eventually boil down to this one condition. It becomes apparent that one should point this out, and thus we have the following theorem.

6.1.3 THEOREM (Markov)

The sequence $X_1, X_2, X_3, \ldots$ obeys the weak law of large numbers if

$$\lim_{n \to \infty} \sigma_{A_n} = 0$$

where $A_n = (1/n)(X_1 + X_2 + \cdots + X_n)$.

Note that as long as $\sigma_{A_n} \to 0$ (henceforth referred to as the Markov condition for short) is assured one need not require that the X_i be independent, or have uniformly bounded variances, and so on. The earlier Theorems 6.1.1 and 6.1.2 can be regarded as corollaries of Theorem 6.1.3.

Now although the Markov condition appears an essential *sufficient* condition to guarantee the weak law of large numbers, it is not a *necessary* condition, as we can see from the following result due to Khintchine (1928).

6.1.4 THEOREM (Khintchine)

The sequence $X_1, X_2, X_3, \ldots$ obeys the weak law of large numbers if
(1) $X_1, X_2, X_3, \ldots$ are independent.
(2) $X_1, X_2, X_3, \ldots$ are all identically distributed as some X.
(3) X has a finite mean but not necessarily a finite variance.

Thus, for example, if X has the density function $f(x) = |x|^{-3}$ for $|x| \geq 1$ and 0 for $|x| < 1$, then $\mu_X = 0$ and $\sigma_X = \infty$ so that $\sigma_{A_n} = \infty$ and the Markov condition cannot possibly hold; and yet according to Khintchine's theorem $X_1, X_2, X_3, \ldots$ will obey the weak law of large numbers if they are independent and have distributions identical to that of X. The proof of Khintchine's theorem employs the so-called "method of truncation" originally due to Markov (1907). This method, which is somewhat intricate, is frequently used in modern probability theory.

PROOF (Optional)
For each n we truncate the first n random variables $X_1, X_2, \ldots, X_n$ as follows. For $k = 1, 2, \ldots, n$, let

$$Y_k = \begin{cases} X_k & \text{whenever } |X_k| \leq n\delta \\ 0 & \text{otherwise} \end{cases}$$

where δ is some arbitrarily chosen positive number. Also let

$$Z_k = \begin{cases} 0 & \text{whenever } |X_k| \leq n\delta \\ X_k & \text{otherwise} \end{cases}$$

Obviously $X_k = Y_k + Z_k$, and Y_k has a finite mean μ_n and a finite variance σ_n^2; in fact

$$\mu_n = \int_{-n\delta}^{n\delta} x P_X(dx) \to \int_{-\infty}^{\infty} x P_X(dx) = \mu$$

and

$$\sigma_n^2 = \int_{-n\delta}^{n\delta} x^2 P_X(dx) - \mu_n^2 \leq n\delta \int_{-n\delta}^{n\delta} |x| P_X(dx) \leq n\delta b$$

where $b = E(|X|)$.

Now let

$$B_n = \frac{Y_1 + Y_2 + \cdots + Y_n}{n}$$

then by Chebyshev's inequality we have

$$P(|B_n - \mu_n| \geq \varepsilon) \leq \frac{\text{var}(B_n)}{\varepsilon^2} = \frac{\sigma_n^2/n}{\varepsilon^2} \leq \frac{\delta b}{\varepsilon^2}$$

Since, for sufficiently large n, $|\mu_n - \mu| < \varepsilon$, we have

$$P(|B_n - \mu| \geq 2\varepsilon) \leq P(|B_n - \mu_n| \geq \varepsilon) \leq \frac{\delta b}{\varepsilon^2}$$

On the other hand let

$$C_n = \frac{Z_1 + Z_2 + \cdots + Z_n}{n}$$

so that $A_n = B_n + C_n$ and $A_n - \mu = B_n - \mu + C_n$, then

$$P(|A_n - \mu| \geq 2\varepsilon) \leq P(|B_n - \mu| \geq 2\varepsilon) + P(C_n \neq 0)$$

since $|A_n - \mu| \geq 2\varepsilon$ implies either $|B_n - \mu| \geq 2\varepsilon$ or else $C_n \neq 0$. But

$$P(C_n \neq 0) \leq \sum_{k=1}^{n} P(Z_k \neq 0)$$

and

$$P(Z_k \neq 0) = \int_{|x| \geq n\delta} P_X(dx) \leq \frac{1}{n\delta} \int_{|x| \geq n\delta} |x| P_X(dx)$$

where the last integral approaches 0 as n approaches ∞ since $E(|x|)$ exists by assumption. Consequently for n sufficiently large, $P(Z_k \neq 0)$ can be made smaller than δ/n and $P(C_n \neq 0)$ smaller than δ. Thus,

$$P(|A_n - \mu| \geq 2\varepsilon) \leq \frac{\delta b}{\varepsilon^2} + \delta$$

for n sufficiently large. But δ is arbitrary, hence

$$\lim_{n \to \infty} P(|A_n - \mu| \geq 2\varepsilon) = 0$$

and the theorem is proved.

A necessary and sufficient condition for a sequence of random variables to obey the weak law of large numbers was found by Gnedenko (1952). It is essentially an improvement on the Markov condition $\sigma_{A_n}^2 = E(A_n - a_n)^2 \to 0$ based on the realization that for some sequence of random variables $E(A_n - a_n)^2$ may not be finite.

6.1.5 THEOREM (Gnedenko)

The sequence $X_1, X_2, X_3, \ldots$ obeys the weak law of large numbers if and only if

$$\lim_{n \to \infty} E\left[\frac{(A_n - a_n)^2}{1 + (A_n - a_n)^2} \right] = 0$$

Note that $E[(A_n - a_n)^2/1 + (A_n - a_n)^2]$ is always finite, in fact less than 1, and also that being less than $E(A_n - a_n)^2 = \sigma_{A_n}^2$ it will approach 0 whenever the latter does, but it may approach 0 even when the latter does not, as must be evident in view of Khintchine's theorem.

PROOF (Optional)

We let F_n denote the distribution function for $A_n - a_n$ and show first that Gnedenko's condition is indeed necessary. Now, assuming $P(|A_n - a_n| \geq \varepsilon) \to 0$ for any given $\varepsilon > 0$, we have

$$P(|A_n - a_n| \geq \varepsilon) = \int_{|x| \geq \varepsilon} dF_n$$

$$\geq \int_{|x| \geq \varepsilon} \frac{x^2}{1 + x^2} dF_n$$

$$= \int \left(\frac{x^2}{1 + x^2} \right) dF_n - \int_{|x| < \varepsilon} \left(\frac{x^2}{1 + x^2} \right) dF_n$$

$$\geq E[(A_n - a_n)^2/1 + (A_n - a_n)^2] - \varepsilon^2$$

As we let $n \to \infty$, we see that

$$\lim_{n \to \infty} E[(A_n - a_n)^2/1 + (A_n - a_n)^2] \leq \varepsilon^2$$

but ε is arbitrary, whence we see that Gnedenko's condition is indeed necessary.

Conversely, if Gnedenko's condition is satisfied, then from

$$P(|A_n - a_n| \geq \varepsilon) = \int_{|x| \geq \varepsilon} dF_n$$

$$\leq \int_{|x| \geq \varepsilon} \frac{1 + \varepsilon^2}{\varepsilon^2} \cdot \frac{x^2}{1 + x^2} dF_n$$

$$\leq \frac{1 + \varepsilon^2}{\varepsilon^2} \int \frac{x^2}{1 + x^2} dF_n$$

$$= \frac{1 + \varepsilon^2}{\varepsilon^2} E[(A_n - a_n)^2 / 1 + (A_n - a_n)^2]$$

we see that $P(|A_n - a_n| \geq \varepsilon) \to 0$. This completes the proof.

Although Gnedenko's condition is both necessary and sufficient, it is, like Markov's condition, not easy to apply in practice since the statement of the condition refers to A_n rather than to the given $X_1, X_2, X_3, \ldots$, and not much is known in general about A_n for a given $X_1, X_2, X_3, \ldots$. In this respect we can see what useful theorems those of Chebyshev and Khintchine really are. Khintchine's theorem is particularly convenient since one does not have to bother with the variance of X.

6.1 PROBLEMS

1. Let random variables X and Y be independent and $\mu_X = 10$, $\sigma_X^2 = 9$, $\mu_Y = 5$, $\sigma_Y^2 = 4$. If $Z = 3X + 2Y$, use the Chebyshev inequality to find a lower estimate for the probability $P(0 \leq Z \leq 70)$. Hint: $P(0 \leq Z \leq 70) \geq P(|Z - 40| \leq 30)$.

2. Derive the following alternative form of the Chebyshev inequality from Proposition 2: If μ and σ are the mean value and the standard deviation of X, then for any positive real number ε we have

$$P(|X - \mu| \geq \varepsilon) \leq \frac{\sigma^2}{\varepsilon^2}$$

Hint: Rewrite ε as $(\varepsilon/\sigma)\sigma$ and apply Proposition 6.1.2.

3. Let $g(x)$ be a monotone increasing positive function [i.e., $0 < g(x_1) < g(x_2)$ for all $x_1 < x_2$] such that $E[g(X)]$ exists for a random variable X. Show that

$$P(X \geq \varepsilon) \leq \frac{E[g(X)]}{g(\varepsilon)}$$

Hint: Imitate the proof of the Chebyshev inequality.

4. Show that $g(x)$ in Problem 6.1.3 need not be monotone increasing and positive over the entire real line but need be so only over the range of X. In particular show that

$$P(|X| \geq \varepsilon) \leq \frac{E(X^2)}{\varepsilon^2}$$

assuming of course that $E(X^2)$ exists. Hint: Let $|X|$ be the random variable in question and $g(x) = x^2$.

5. Let k be a positive even integer and μ a real number such that $E(|X - \mu|^k)$ exists for the random variable X. Show that

$$P(|X - \mu| \geq \varepsilon) \leq \frac{E(|X - \mu|^k)}{\varepsilon^k}$$

Note that for $k = 2$ we have the Chebyshev inequality in Problem 6.1.2. Hint: Let $|X - \mu|$ be the random variable in question and $g(x) = x^k$.

6. Optional. Show that Markov's law of large numbers (Theorem 6.1.3) implies Chebyshev's law of large numbers (Theorem 6.1.2), also that the latter implies Bernoulli's law of large numbers (Theorem 6.1.1).

7. If X_1, X_2, X_3, ... are mutually independent and each X_n for $n = 1, 2, 3, ...$ has a symmetric two-point distribution $P(X_n = n^{1/3}) = P(X_n = -n^{1/3})$, show that they obey the weak law of large numbers. Hint: Use Theorem 6.1.3, also that $1^\alpha + 2^\alpha + \cdots + n^\alpha \leq \int_0^{n+1} x^\alpha \, dx$ for $\alpha > 0$.

8. If X_1, X_2, X_3, ... are mutually independent and are all identically distributed with Cauchy density function $f_X(x) = 1/[\pi(1 + x^2)]$, show that they do not obey the weak law of large numbers. Hint: Use Section 3.6, Example 5, to show that all sample averages A_n have distributions identical to that of X.

9. If A_n is the sample average of a random variable X with a finite mean μ, then the probability mass of A_n will cluster around μ (see Khintchine's theorem). Thus if f is a continuous function, the probability mass of $f(A_n)$ ought to cluster around $f(\mu)$ so that we may expect $E[f(A_n)]$ to be close to $f(\mu)$. Prove that

$$\lim_{n \to \infty} E[f(A_n)] = f(\mu)$$

if f is a bounded continuous function.

10. Probabilistic view of Weierstrass approximation theorem. Let f be a continuous function defined for $0 \leq p \leq 1$. According to Weierstrass a continuous function $f(p)$ on $[0, 1]$ can be uniformly approximated to within an arbitrary $\varepsilon > 0$ by a polynomial in p. Prove this assertion by identifying p with the mean value of a random variable X with a two-point distribution $P(X = 1) = p$ and $P(X = 0) = 1 - p$ and writing out $E[f(A_n)]$ as an approximation of $f(p)$ where A_n is the sample average of X as in Problem 6.1.9.

6.2 THE STRONG LAW OF LARGE NUMBERS

The weak law of large numbers (say Bernoulli's version, Theorem 6.1.1) as a mathematical formulation of what is empirically felt by experimenters (say of coin tossing) is undoubtedly one of the important achievements in the early history of probability theory. However, it does not really say quite enough. One thing it fails to say is that if we keep tossing a coin, the average number of heads (i.e., the ratio of heads to tosses) will almost certainly (i.e., with probability 1) approach to some fixed number. For while logically it is possible (and the weak law of large numbers certainly does not preclude this) that a coin keeps falling heads and tails in such a way that the average number of heads fluctuates forever or approaches motley unlikely numbers (such as 0 or 1), somehow we sense that this is quite unlikely. We feel somehow that if we keep tossing the coin, the average number of heads will almost certainly approach some unknown fixed number (maybe $\frac{1}{2}$ for an ideal coin). This empirical feeling is given a precise mathematical formulation known as the strong law of large numbers, which we can prove as a mathematical theorem in the context of a mathematical model.

6.2.1 THEOREM (A Strong Law of Large Numbers)

If A_n is the sample average of X whose mean value μ and *fourth central moment* $\tau^4 = E[(X - \mu)^4]$ are finite, then

$$P(\lim_{n \to \infty} A_n = \mu) = 1$$

In plain words, the sample average will tend to a fixed constant with probability 1. If X is the random variable originally representing the coin (so that $P_X(1) = p$ and $P_X(0) = 1 - p$ and consequently $\mu_X = p$), then A_n represents the ratio of the number of heads to the number of tosses at the completion of the first n tosses, and the theorem may be interpreted as saying that this ratio will tend to a fixed constant almost certainly. Although it is easy to say that within our mathematical model this constant is none other than the mean value of X (namely p), outside of the mathematical model this constant can only be guessed by tossing the coin a large number of times. In other words, there is no definite way of actually determining the probability of a given coin falling heads. However, if we choose a probability value based on a large number of tosses of this coin, the resulting mathematical model will perhaps not be too far off the reality whatever this reality is. In this connection we must point out that reality is never definable, but can only be described (and not necessarily uniquely, either) in terms of models, mathematical or otherwise.

Theorem 6.2.1 is essentially due to Borel (1909). The proof is accomplished in two steps. Since these steps are somewhat complicated the reader may wish to go on to page 230.

We begin with establishing the following analogs of Propositions 6.1.1. and 6.1.2.

6.2.1 PROPOSITION

If X has a finite mean value μ and a finite fourth central moment $\tau^4 = E[(X - \mu)^4]$ (hence also all the lower central moments including the variance σ^2), then the sample average A_n of X also has a finite mean value $\mu(A_n)$ and a finite fourth central moment $\tau^4(A_n)$, in fact

$$\mu(A_n) = \mu$$

$$\tau^4(A_n) = \frac{\tau^4 + 3(n-1)\sigma^4}{n^3}$$

PROOF

We have already calculated $\mu(A_n) = \mu$ in the preceding section (Proposition 6.1.1). To find $\tau^4(A_n)$ we first calculate $\tau^4(S_n)$ where $S_n = X_1 + X_2 + \cdots + X_n = nA_n$.

$$\tau^4(S_n) = E(S_n - \mu_{S_n})^4 = E(X_1 + X_2 + \cdots + X_n - n\mu)^4$$
$$= E[(X_1 - \mu) + (X_2 - \mu) + \cdots + (X_n - \mu)]^4$$
$$= E\left[\sum_i (X_i - \mu)^4 + \sum_{i<j} C_2^4 (X_i - \mu)^2 (X_j - \mu)^2\right]$$

where all terms of the form $(X_i - \mu)(X_j - \mu)^3$ are dropped since $X_i - \mu$ and $(X_j - \mu)^3$ are independent and $E(X_i - \mu) = 0$. Consequently,

$$\tau^4(S_n) = n\tau^4 + \frac{n(n-1)}{2} \cdot 6\,\sigma^2 \cdot \sigma^2$$

Therefore

$$\tau^4(A_n) = E(A_n - \mu_{A_n})^4 = \frac{1}{n^4} E(S_n - \mu_{S_n})^4$$
$$= \frac{\tau^4(S_n)}{n^4} = \frac{\tau^4 + 3(n-1)\sigma^4}{n^3}$$

Proposition 6.2.1 will be made useful by the following generalized Chebyshev inequality just as Proposition 6.1.1 was.

6.2.2 PROPOSITION (Generalized Chebyshev Inequality)

If μ and τ^k are respectively the mean value and the kth central moment of X, then for any real positive number n we have

$$P(|X - \mu| \geq n\tau_k) \leq \frac{1}{n^k}$$

provided that k is an even integer and where τ_k is the kth root of the kth central moment, $\tau_k = \sqrt[k]{\tau^k}$ (call it the kth order *standard deviation* if you like).

PROOF

In $\tau^k = E[(X - \mu)^k]$ let $Y = (X - \mu)^k$ and let

$$Z = \begin{cases} (n\tau_k)^k & \text{whenever } Y \geq (n\tau_k)^k \\ 0 & \text{whenever } Y < (n\tau_k)^k \end{cases}$$

then

$$Z \leq Y \quad \text{and} \quad E(Z) \leq E(Y) = \tau^k$$

but

$$E(Z) = (n\tau_k)^k \cdot P(Y \geq (n\tau_k)^k)$$
$$= n^k\tau^k P(|X - \mu| \geq n\tau_k)$$

from which follows

$$P(|X - \mu| \geq n\tau_k) \leq \frac{1}{n^k}$$

The following alternative form of the generalized Chebyshev inequality may prove handy, as in Lemma 6.2.1 to follow.

6.2.3 PROPOSITION (Alternative Form of Chebyshev Inequality)

For any even integer k if $E[(X - \mu)^k] = \tau^k$ exists, then for any real number $\varepsilon > 0$

$$P(|X - \mu| \geq \varepsilon) \leq \frac{\tau^k}{\varepsilon^k}$$

In particular, for $k = 2$, we have

$$P(|X - \mu| \geq \varepsilon) \leq \frac{\sigma^2}{\varepsilon^2}$$

PROOF

In view of Proposition 6.2.2 we have

$$P(|X - \mu| \leq \varepsilon) = P(|X - \mu| \geq (\varepsilon/\tau_k)\tau_k)$$
$$\leq \frac{1}{(\varepsilon/\tau_k)^k} = \frac{\tau^k}{\varepsilon^k}$$

Propositions 6.2.1 and 6.2.3 lead to the following lemma, which can be regarded as a strengthening of Theorem 6.1.1, which says essentially that $P(|A_n - \mu| \geq \varepsilon)$ is dominated by $1/n$.

6.2.1 LEMMA

If A_n is the sample average of X, whose mean value μ and fourth central moment τ^4 are both finite, then

$$P(|A_n - \mu| \geq \varepsilon) \leq \frac{\tau^4 + 3\sigma^4}{\varepsilon^4} \frac{1}{n^2}$$

for any $\varepsilon > 0$ however small.

PROOF

Using the alternative form of the Chebyshev inequality (Proposition 6.2.3), we have

$$P(|A_n - \mu| \geq \varepsilon) \leq \frac{\tau^4(A_n)}{\varepsilon^4}$$

But by Proposition 6.2.1

$$\tau^4(A_n) = \frac{\tau^4 + 3(n-1)\sigma^4}{n^3} < \frac{n\tau^4 + 3n\sigma^4}{n^3} = \frac{\tau^4 + 3\sigma^4}{n^2}$$

Consequently,

$$P(|A_n - \mu| \geq \varepsilon) \leq \frac{\tau^4 + 3\sigma^4}{\varepsilon^4} \frac{1}{n^2}$$

Thus, being dominated by $1/n^2$, $P(|A_n - \mu| \geq \varepsilon)$ will approach 0 even faster than we thought in Theorem 6.1.1. In this connection it is important to point out that there is an essential difference between a sequence of numbers x_n dominated by $1/n$ and another y_n dominated by $1/n^2$. Although in both cases the terms of the sequence approach 0 so that both x_n and y_n are small when n is large, with the former sequence these small numbers may add up to ∞ (consider $x_n = 1/n$ for example) while with the latter sequence this cannot happen. This is due to the fact that while $\sum (1/n)$ is divergent, $\sum (1/n^2)$ is convergent.

We are ready to establish the strong law of large numbers except for one final important detail concerning our mathematical model. In writing $P(|A_n - \mu| \geq \varepsilon)$ we have taken P for granted. But are we talking about the same P for all n? And if so, what is the underlying probability space with which this P is associated? Let us briefly sketch the underlying $(S, \mathscr{S}, P)$ that ought to have been in the back of our mind. Recall that we started out with infinitely many copies of X: $X_1, X_2, X_3, \ldots$ and assumed that they were independent. This means we took for granted a certain probability space $(S, \mathscr{S}, P)$ on which $X_1, X_2, X_3, \ldots$ are defined and are mutually independent in the sense that any finite subclass of X_i is stochastically independent. What is this $(S, \mathscr{S}, P)$ really?

First of all, S consists of all infinite sequences of real numbers

$$S = R^\infty = \{\langle x_1, x_2, x_3, \ldots \rangle : x_i \in R\}$$

on which X_i is a random variable such that

$$X_i(\langle x_1, x_2, x_3, \ldots \rangle) = x_i$$

and in which the event that X_i takes a value in a Borel set B is represented by

$$B_i = \{\langle x_1, x_2, x_3, \ldots \rangle : x_i \in B\}$$

The family $\mathscr{S}$ of events is then defined as the smallest sigma algebra of subsets of R^∞ including all possible B_i as members.

Finally, the probability measure P is defined (and this can be shown to be possible in one and only one way: see Minibibliography: Royden) in such a way that

$P(B_i) =$ the probability that the given random variable X takes a value in B

for every i and every Borel set B and that any finite class of events $B_1, B_2, \ldots, B_n$ are independent. This completes the description of our underlying probability space $(S, \mathscr{S}, P)$. Needless to say, for each n, $A_n = (1/n)(X_1 + X_2 + \cdots + X_n)$ is a random variable on this probability space and the probability $P(\lim_{n \to \infty} A_n = \mu)$ is simply the probability of the event

$$E = \left\{ \omega \in S \colon \lim_{n \to \infty} A_n(\omega) = \mu \right\}$$

PROOF OF THEOREM 6.2.1

Denoting the underlying probability space for $X_1, X_2, \ldots$, and hence for all A_n by $(S, \mathscr{S}, P)$, and letting

$$E = \left\{ \omega \in S \colon \lim_{n \to \infty} |A_n(\omega) - \mu| = 0 \right\}$$

we want to show

$$P(E) = 1$$

To this end we shall show

$$P(D) = 0$$

where $D = S - E$.

Now handle D as follows:

$$D = \bigcup_{k=1}^{\infty} D_{1/k}$$

where

$$D_{1/k} = \left\{ \omega \in S \colon \lim_{n \to \infty} |A_n(\omega) - \mu| \geqq \frac{1}{k} \right\}$$

in which $\lim_{n \to \infty} |A_n(\omega) - \mu| \geqq 1/k$ is interpreted to mean $|A_n(\omega) - \mu| \geqq 1/k$ for infinitely many n since in truth we do not really know whether $\lim_{n \to \infty} |A_n(\omega) - \mu|$ exists for each ω. The theorem will have been proved if we can only show

$$P(D_{1/k}) = 0 \qquad \text{for} \quad k = 1, 2, 3, \ldots$$

since $P(D) \leqq \sum_k P(D_{1/k})$.

But now

$$D_{1/k} \subset D_{1/k, n} \cup D_{1/k, n+1} \cup \cdots$$

$$= \bigcup_{m=0}^{\infty} D_{1/k, n+m}$$

for any n however large, where

$$D_{1/k,\,n} = \left\{ \omega \in S \colon |A_n(\omega) - \mu| \geq \frac{1}{k} \right\}$$

Recall however that according to Lemma 6.2.1, $P(D_{1/k,\,n})$ is dominated by $1/n^2$ so that $P(D_{1/k}) \leq P(D_{1/k,\,n}) + P(D_{1/k,\,n+1}) + \cdots$ and the right-hand side of this inequality is made arbitrarily small by taking n sufficiently large. Therefore, $P(D_{1/k}) = 0$, and this completes the proof.

The following development toward greater generality parallels that of the weak law of large numbers in Section 6.1.

6.2.1 DEFINITION

A sequence of random variables $X_1, X_2, X_3, \ldots$ is said to obey the strong law of large numbers if

$$P\left(\lim_{n \to \infty} |A_n - a_n| = 0 \right) = 1$$

where

$$A_n = \frac{1}{n}(X_1 + X_2 + \cdots + X_n)$$

and

$$a_n = \frac{1}{n}(\mu_{X_1} + \mu_{X_2} + \cdots + \mu_{X_n})$$

In words, the *random average* A_n approaches the *means average* a_n with probability 1. According to this definition Theorem 6.2.1 simply says that a set of sufficient conditions for $X_1, X_2, X_3, \ldots$ to obey the strong law of large numbers is

(1) $X_1, X_2, X_3, \ldots$ are independent.
(2) $X_1, X_2, X_3, \ldots$ are identically distributed as some X.
(3) X (and hence each X_i) has a finite fourth (and hence all the lower) moments.

Weakening of above conditions leads to generalizations of Theorem 6.2.1. We state some of these (cf. Minibibliography: Gnedenko).

6.2.2 THEOREM (Kolmogorov)

The sequence $X_1, X_2, X_3, \ldots$ obeys the strong law of large numbers if
(1) $X_1, X_2, X_3, \ldots$ are independent.
(2) $\displaystyle\sum_{n=1}^{\infty} \frac{\text{var}(X_n)}{n^2} < \infty.$

6.2.3 THEOREM (Kolmogorov)

The sequence $X_1, X_2, X_3, \ldots$ obeys the strong law of large numbers if
(1) X_1, X_2, X_3, are ... independent.
(2) $X_1, X_2, X_3, \ldots$ are all identically distributed as some X.
(3) X has a finite mean.

This last theorem of Kolmogorov (1930) bears a striking resemblance to the earlier theorem of Khintchine (Theorem 6.1.4) and again impresses us as an extremely neat result. Actually Kolmogorov has shown more than is stated in the above theorem; he claims in fact that for a sequence of independent identically distributed random variables the existence of finite means is a necessary and sufficient condition for the strong law of large numbers to hold.

Both the weak and the strong laws of large numbers assert that a sequence of random variables in some fashion approaches a constant (a special kind of random variable). This point of view suggests that we consider various types of convergence for sequences of random variables and investigate which types will imply (hence can be regarded as stronger than) which other types. Therefore, let us consider the following definition.

6.2.2 DEFINITION

A sequence of random variables $X_1, X_2, \ldots$ defined on a probability space $(S, \mathscr{S}, P)$ is said to *converge almost surely* to a random variable X_0 if for almost all $\omega \in S$, $X_n(\omega)$ will approach $X_0(\omega)$, that is:

$$P\left\{\omega: \lim_{n \to \infty} |X_n(\omega) - X_0(\omega)| = 0\right\} = 1$$

Thus the strong law of large numbers says that under certain suitable assumptions the random averages A_n converge almost certainly to a constant.

6.2.3 DEFINITION

A sequence of random variables $X_1, X_2, \ldots$ defined on a probability space $(S, \mathscr{S}, P)$ is said to *converge in probability* to a random variable X_0 if for any $\varepsilon > 0$

$$\lim_{n \to \infty} P\{\omega: |X_n(\omega) - X_0(\omega)| \leqq \varepsilon\} = 1$$

Thus the weak law of large numbers says that under certain suitable assumptions the random averages A_n converge in probability to a constant.

6.2.4 DEFINITION

A sequence of random variables $X_1, X_2, \ldots$ defined on a probability space $(S, \mathscr{S}, P)$ is said to *converge in the mean* to a random variable X_0 if

$$\lim_{n \to \infty} E(X_n - X_0)^2 = 0$$

The preceding types of convergence together with another type called *convergence in distribution* (see Definition 6.4.3) constitute the four well-known types of convergence in probability theory. The pattern of implication of these four types of convergence is as follows (cf. Minibibliography: Loève):

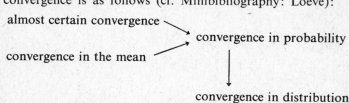

Thus, in as much as almost certain convergence implies convergence in probability, Bernoulli's weak law of large numbers (Theorem 6.1.1) may be regarded essentially as a corollary to Borel's strong law of large numbers (Theorem 6.2.1), and likewise Khintchine's weak law of large numbers (Theorem 6.1.4) as a corollary to Kolmogorov's strong law of large numbers (Theorem 6.2.3).

6.2 PROBLEMS

1. Given a sequence of events D_1, D_2, D_3, ..., show the following.
   a. If an outcome ω belongs to at most finitely many of them (say, the first dozen), then

$$\omega \notin \bigcup_{m=0}^{\infty} D_{n+m}$$

for some sufficiently large n (say 13).
   b. If an outcome ω belongs to infinitely many of them (say, every other one), then

$$\omega \in \bigcup_{m=0}^{\infty} D_{n+m}$$

for any n however large; consequently

$$\omega \in \bigcap_{n=1}^{\infty} \bigcup_{m=0}^{\infty} D_{n+m}$$

2. Given a sequence of sets D_1, D_2, D_3, ... the set of points belonging to infinitely many of them is called *lim sup D_n*. Show

$$\limsup D_n = \bigcap_{n=1}^{\infty} \bigcup_{m=0}^{\infty} D_{n+m}$$

On the other hand, the set of points belonging to almost all (all but finitely many) of them is called *lim inf D_n*. Clearly lim inf $D_n \subset$ lim sup D_n. Show that

$$\liminf D_n = \bigcup_{n=1}^{\infty} \bigcap_{m=0}^{\infty} D_{n+m}$$

3. Given a sequence of events $D_1, D_2, D_3, \ldots$ in a probability space $(S, \mathscr{S}, P)$ an outcome ω either belongs to at most finitely many D's or else belongs to infinitely many D's. The well-known *Borel-Cantelli lemma* says:

a. If $\sum_{n=1}^{\infty} P(D_n) < \infty$, then with probability 1 an outcome belongs to at most finitely many D's, that is:

$$P(\limsup D_n) = 0$$

b. If $\sum_{n=1}^{\infty} P(D_n) = \infty$ and D_n are mutually independent, then with probability 1 an outcome belongs to infinitely many D's, that is:

$$P(\limsup D_n) = 1$$

Prove (a) and think about the plausibility of (b). Hint: (a) Since $\sum_{n=1}^{\infty} P(D_n)$ converges, given an arbitrarily small $\varepsilon > 0$ there exists an n such that $\sum_{m=0}^{\infty} P(D_{n+m}) < \varepsilon$ so that $P(\bigcup_{m=0}^{\infty} D_{n+m}) < \varepsilon$, but $P(\limsup D_n) \leq P(\bigcup_{m=0}^{\infty} D_{n+m})$ for any n.

4. Proof of Borel's strong law of large numbers via Borel-Cantelli lemma. If A_n is the sample average of X whose mean value μ and fourth central moment τ^4 are finite (see Lemma 6.2.1), then for any $\varepsilon > 0$

$$P(|A_n - \mu| \geq \varepsilon) \leq \frac{c}{n^2}$$

where c is a constant having nothing to do with n.

a. Let D_n be the event $\{\omega \in S : |A_n(\omega) - \mu| \geq \varepsilon\}$ and show that $\sum_{n=1}^{\infty} P(D_n) < \infty$. Then apply the Borel-Cantelli lemma (a) of Problem 6.2.3 to show that

$$P(\limsup D_n) = 0$$

b. Now since the event $\limsup D_n$ depends on ε, let

$$E_\varepsilon = \limsup D_n$$

and interpret the event $S - E_\varepsilon$ as well as the probability $P(S - E_\varepsilon) = 1$.

c. If $\limsup |A_n - \mu| < \varepsilon$ with probability 1 for any ε, show that $\limsup |A_n - \mu| = 0$ with probability 1, and hence $\lim |A_n - \mu| = 0$ with probability 1.

5. Optional. Construct a sequence of functions f_n on the interval $[0, 1]$ converging pointwise on $[0, 1]$ to $f_0 \equiv 0$ but not converging in the mean to f_0. Likewise but conversely, construct f_n converging to f_0 in the mean but not anywhere on $[0, 1]$.

6.3 CHARACTERISTIC FUNCTIONS

The distribution of a given random variable X is completely described by its distribution function F_X. Distribution functions are real-valued functions of a real variable; they are useful as media for determining probability distributions. However, their usefulness is curtailed by their unwieldy behavior under convolution; difficulty in convoluting two distribution functions makes them undesirable as a tool for investigating distributions of sums of independent random variables. The question therefore arises: What numerical function is there that will characterize the probability distribution and at the same time have a simple behavior under convolution? In other words, for any given random variable X can we assign a certain function of a real variable ϕ_X such that if X and Y have distinct distributions, ϕ_X and ϕ_Y will be distinct, and furthermore, such that if X and Y are independent ϕ_{X+Y} can be computed from ϕ_X and ϕ_Y by some simple mathematical operation? The affirmative answer is found in the so-called characteristic functions to be introduced in this section. These functions may be regarded as evolving from a certain prototype known as moment-generating functions. Much about characteristic functions is revealed by studying moment-generating functions, which are somewhat more intuitive.

6.3.1 DEFINITION

The moment-generating function of a random variable X is a real-valued function of a real variable given by

$$M_X(t) = E(e^{tX})$$

for all t such that the expected value of e^{tX} is finite.

Note that for each fixed t, e^{tX} is a random variable derived from X by composing X with the Borel function g where $g(x) = e^{tx}$. Now why should the knowledge about $E[g(X)]$ for various g be important for us? First of all, it is easy to see that if we know the values of $E[g(X)]$ for all conceivable g, in particular the indicator functions

$$I_a(x) = \begin{cases} 1 & \text{for} \quad x \leq a \\ 0 & \text{for} \quad x > a \end{cases}$$

then from

$$\begin{aligned} E[I_a(X)] &= 1 \cdot P_X(-\infty, a] + 0 \cdot P_X(a, \infty) \\ &= P_X(-\infty, a] \\ &= F_X(a) \end{aligned}$$

we know also the values of the distribution function F_X, but knowing F_X is tantamount to knowing the distribution P_X. Consequently, if we know the values of $E[g(X)]$ for all conceivable g, we certainly know the distribution of X. But then perhaps we need not know $E[g(X)]$ for all g in order to know P_X, for in fact we have just seen that the knowledge of $E[I_a(X)]$ for all a suffices. Now the moment-generating function $M_X(t)$ embodies the knowledge of $E[e^{tX}]$ for all t. Shouldn't $M_X(t)$ then tell us quite a bit about the distribution of X? It does indeed. For example it may tell us all about the moments of X (hence the name moment-generating function). And it behaves well under convolution. We shall now elaborate on these with propositions and examples.

6.3.1 PROPOSITION

If X has a moment-generating function $M_X(t)$ defined in some neighborhood of $t = 0$, then the nth moment of X can be obtained by taking the nth derivative of $M_X(t)$ and setting $t = 0$,

$$M_X^{(n)}(0) = E(X^n)$$

Note that for any X, $M_X(t)$ is always defined at least at $t = 0$ since $M_X(0) = E(e^{0 \cdot X}) = 1$, but for some X, $M_X(t)$ may not be defined anywhere except at $t = 0$ (see Problem 6.3.1).

PROOF

For heuristic reasons let us first assume that X has a density function $f_X(x)$ and proceed somewhat mechanically as follows. For each t in the given neighborhood of $t = 0$, we have

$$M_X(t) = E(e^{tX}) = \int_{-\infty}^{\infty} e^{tx} f_X(x)\, dx$$

$$\doteq \int_{-\infty}^{\infty} \left(1 + tx + \frac{t^2 x^2}{2!} + \cdots \right) f_X(x)\, dx$$

$$= \int_{-\infty}^{\infty} f_X(x)\, dx + \frac{t}{1!} \int_{-\infty}^{\infty} x f_X(x)\, dx + \cdots$$

$$= 1 + E(X)\frac{t}{1!} + E(X^2)\frac{t^2}{2!} + \cdots$$

Differentiating with respect to t, we have

$$M_X'(t) = E(X) + E(X^2)\frac{t}{1!} + E(X^3)\frac{t^2}{2!} + \cdots$$

Setting $t = 0$, we obtain

$$M_X'(0) = E(X)$$

Differentiating once more, we have

$$M_X''(t) = E(X^2) + E(X^3)\frac{t}{1!} + \cdots$$

Setting $t = 0$ again, we obtain

$$M_X''(0) = E(X^2)$$

Continuing thus, we obtain in general

$$M_X^{(n)}(0) = E(X^n)$$

We now give up the assumption that X has a density function and proceed directly as follows. We have

$$M_X(t) = E(e^{tX}) = \int_S e^{tX(\omega)} P(d\omega)$$

for all t for which the integral exists. Consequently we have

$$M_X^{(n)}(t) = \frac{d^n}{dt^n} \int_S e^{tX(\omega)} P(d\omega)$$

$$= \int_S \frac{d^n}{dt^n} e^{tX(\omega)} P(d\omega)$$

for all n for which the last integral exists. Thus we have

$$M_X^{(n)}(t) = \int_S X^n(\omega) e^{tX(\omega)} P(d\omega)$$

Setting $t = 0$, we obtain

$$M_X^{(n)}(0) = \int_S X^n(\omega) P(d\omega) = E(X^n)$$

Example 1 Find the moment-generating function $M_X(t)$ of a random variable X having the standard normal distribution

$$f_X(x) = \frac{1}{\sqrt{2\pi}} e^{-x^2/2}$$

and calculate the first three moments of X from M_X.

Solution For any real value t we have

$$M_X(t) = \frac{1}{\sqrt{2\pi}} \int_{-\infty}^{\infty} e^{tx} e^{-x^2/2} \, dx$$

Combining the exponents and completing the square, that is:

$$tx - \tfrac{1}{2}x^2 = -\tfrac{1}{2}(x - t)^2 + \tfrac{1}{2}t^2$$

we see

$$M_X(t) = e^{t^2/2} \left[\frac{1}{\sqrt{2\pi}} \int_{-\infty}^{\infty} e^{-(x-t)^2/2} \, dx \right]$$

By a simple change of variable the integral is reduced to the integral of the standard normal density function, which is 1. Hence

$$M_X(t) = e^{t^2/2}$$

Now differentiating $M_X(t)$ with respect to t, we obtain

$$M_X'(t) = e^{t^2/2}t$$

so that $E(X) = M_X'(0) = 0$, which we know already. Differentiating again, we have

$$M_X''(t) = e^{t^2/2}t^2 + e^{t^2/2}$$

so that $E(X^2) = M_X''(0) = 1$. Differentiating once more,

$$M_X'''(t) = e^{t^2/2}t^3 + 3te^{t^2/2}$$

so that $E(X_3) = M_X'''(0) = 0$.

Example 2 Find the moment-generating function of a (μ, σ) normal variable Y from that of the standard normal variable X.

Solution If Y is (μ, σ) normal, then we have seen that $(Y - \mu)/\sigma = X$ is standard normal. In other words $Y = \sigma X + \mu$ is a simple linear transform of X. Now, in general, if $Y = aX + b$ is a linear transform of X whose moment-generating function $M_X(t)$ is already known, then we can calculate $M_Y(t)$ as follows:

$$
\begin{aligned}
M_Y(t) = E(e^{tY}) &= E[e^{t(aX+b)}] \\
&= E(e^{tb} \cdot e^{taX}) \\
&= e^{tb}E(e^{taX}) \\
&= e^{tb}M_X(at)
\end{aligned}
$$

Consequently now if $M_X(t) = e^{t^2/2}$, and $Y = \sigma X + \mu$, then

$$M_Y(t) = e^{\mu t}e^{(\sigma t)^2/2} = e^{\mu t + \sigma^2 t^2/2}$$

We recall that whenever the sum $X + Y$ of two independent random variables is formed, practically "anything" related to $X + Y$ is called the convolution of these same things related to X and Y. For example, the density function f_{X+Y} is called the convolution of the density functions f_X and f_Y; briefly we wrote

$$f_{X+Y} = f_X * f_Y$$

if X and Y are independent. Unfortunately, however, $f_X * f_Y$ is in general not easy to calculate from f_X and f_Y. We have likewise

$$P_{X+Y} = P_X * P_Y$$
$$F_{X+Y} = F_X * F_Y$$

for independent X and Y, but then again the calculations of $P_X * P_Y$ and $F_X * F_Y$ may be difficult. Under these circumstances we welcome the following proposition.

6.3.2 PROPOSITION

If X and Y are independent and have moment-generating functions $M_X(t)$ and $M_Y(t)$ in some neighborhood around $t = 0$, then $X + Y$ also has a moment-generating function $M_{X+Y}(t)$ in the neighborhood of $t = 0$, and in fact

$$M_{X+Y}(t) = M_X(t) \cdot M_Y(t)$$

In other words, the convolution of M_X and M_Y may be found by simply multiplying these two functions; thus we calculate $M_X * M_Y$ from

$$(M_X * M_Y)(t) = M_X(t) \cdot M_Y(t)$$

PROOF
Following the definition and remembering that X and Y are independent, we have

$$M_{X+Y}(t) = E[e^{t(X+Y)}] = E(e^{tX}) \cdot E(e^{tY})$$
$$= M_X(t) \cdot M_Y(t)$$

Example 3 Find the moment-generating function of X having a Poisson distribution

$$P_X(k) = e^{-\mu} \frac{\mu^k}{k!} \qquad \text{for} \quad k = 0, 1, 2, \ldots$$

Find also the moment-generating function of $X + Y$ if X and Y are independent and both have Poisson distributions, say

$$P_Y(k) = e^{-\lambda} \frac{\lambda^k}{k!} \qquad \text{for} \quad k = 0, 1, 2, \ldots$$

and P_X as given above.

Solution For any real t we have

$$M_X(t) = \sum_{k=0}^{\infty} e^{tk} \cdot e^{-\mu} \frac{\mu^k}{k!}$$

$$= e^{-\mu} \sum_{k=0}^{\infty} \frac{(\mu e^t)^k}{k!}$$

$$= e^{-\mu} \cdot e^{\mu e^t}$$

$$= e^{\mu(e^t - 1)}$$

Now if Y has a λ-Poisson distribution, then likewise we have

$$M_Y(t) = e^{\lambda(e^t - 1)}$$

Hence, according to Proposition 6.3.2, we have

$$M_{X+Y}(t) = e^{\mu(e^t - 1)} \cdot e^{\lambda(e^t - 1)}$$
$$= e^{(\mu + \lambda)(e^t - 1)}.$$

This result is significant in that the moment-generating function of $X + Y$ is that of a random variable having $(\mu + \lambda)$ Poisson distribution. We suspect therefore that the sum of Poisson variables is again a Poisson variable. This is indeed the case because it can be shown that if two random variables have identical moment-generating functions in some neighborhood of $t = 0$, then they must have the same distribution. However, we shall also confirm this with the aid of characteristic functions to be introduced presently.

Now what is wrong with the moment-generating functions that we must seek improvements on them? One must be aware by now that despite all the nice properties, the moment-generating functions as a mathematical tool does have a serious defect; namely for some random variables with "outgoing" distributions the moment-generating functions may not be defined anywhere except at $t = 0$. When this is the case (see Problem 6.3.5), the moment-generating functions no longer serve to distinguish (or characterize) distinct distributions. This is where the characteristic functions come in.

6.3.2 DEFINITION

The characteristic function of a random variable X is a complex-valued function of a real variable given by

$$\phi_X(t) = E(e^{itX})$$

Note that i is the imaginary number $\sqrt{-1}$ and

$$e^{i\theta} = \cos\theta + i\sin\theta$$

so that

$$E(e^{itX}) = E[\cos tX + i\sin tX]$$
$$= E(\cos tX) + i\,E(\sin tX)$$

Since $\cos tX$ and $\sin tX$ are bounded random variables for each t, both $E(\cos tX)$ and $E(\sin tX)$ are finite and hence $E(e^{itX})$ is defined for each t, and all this for any X no matter how widespread its distribution is.

What we are about to state amounts to this: For any given X the knowledge of $E(\cos tX)$ and $E(\sin tX)$ for all t suffices to determine the distribution P_X and whence further the values $E[g(X)]$ for all Borel functions g. In other words, ϕ_X completely characterizes P_X.

6.3.3 PROPOSITION

Given a random variable X if its characteristic function ϕ_X is known, then its probability distribution P_X can be determined from ϕ_X by the following "inversion formula":

$$P_X(x_1, x_2) + \tfrac{1}{2}P_X(x_1) + \tfrac{1}{2}P_X(x_2) = \lim_{T \to \infty} \frac{1}{2\pi} \int_{-T}^{T} \frac{e^{-itx_1} - e^{-itx_2}}{it} \phi_X(t)\,dt$$

for any $x_1 < x_2$. The integrand at $t = 0$ is defined by the limit as t approaches 0.

The proof of this proposition can be found in advanced probability textbooks (e.g., see Minibibliography: Chung). What is important for us here is the significance of the inversion formula, which says that the amount of probability mass on any interval (x_1, x_2) plus half of the mass on the end points x_1 and x_2 is determinable from the characteristic function. Denoting this amount by $\tilde{P}_X(x_1, x_2)$, it is not too hard to see that for the closed interval $[x_1, x_2]$ we have

$$P_X[x_1, x_2] = \lim_{n \to \infty} \tilde{P}_X\left(x_1 - \frac{1}{n}, x_2 + \frac{1}{n}\right)$$

and further that for the half-closed interval $(x_1, x_2]$ we have

$$P_X(x_1, x_2] = \lim_{n \to \infty} P_X\left[x_1 + \frac{1}{n}, x_2\right]$$

Now since $F_X(x) = P_X(-\infty, x]$ and $(-\infty, x]$ is the countable union of half-closed intervals of the type just considered, $F_X(x)$ is ultimately determinable from $\tilde{P}_X$ and hence from ϕ_X. For example, if x is a real number lying between the integer N and $N + 1$, then

$$P_X(-\infty, x] = \bigcup_{n=-\infty}^{N} P_X(n - 1, n] + P_X(N, x]$$

The point of all this is that ϕ_X determines F_X, which as we know determines P_X; consequently the characteristic function ϕ_X determines (or characterizes, and hence the name *characteristic function*) the probability distribution P_X.

Characteristic functions eliminate the major defect of moment-generating functions while retaining most of the nice properties. Consequently, we have the following two propositions, the first of which can be proved routinely while the second cannot be proved with the means at our disposal (see Minibibliography: Chung).

6.3.4 PROPOSITION

If X and Y are independent random variables, we have

$$\phi_{X+Y}(t) = \phi_X(t)\phi_Y(t)$$

If Y is related to X by $Y = aX + b$, then we have

$$\phi_Y(t) = e^{ibt}\phi_X(at)$$

We see that characteristic functions behave just like moment-generating functions under convolution and also under linear distortions except the i in e^{ibt}. The moment-generating property cannot hold for all characteristic functions, however, since characteristic functions are defined even for random variables having no moments [see Problem 6.3.7g]. We have nevertheless just about the best result possible under the circumstances as stated in the following proposition.

6.3.5 PROPOSITION

If a random variable X has a finite absolute moment of order k

$$E(|X|^k) < \infty$$

where k is a positive integer, then the characteristic function ϕ_X has a continuous derivative of order k given by

$$\phi_X^{(k)}(t) = E[(iX)^k e^{itX}]$$

so that if we set $t = 0$, we obtain

$$\phi_X^{(k)}(0) = i^k E(X^k)$$

Conversely, if ϕ_X has a finite derivative of even order k at $t = 0$, then X has a finite moment $E(X^k)$ and hence also finite $E(X^j)$ for all $j < k$.

Example 4 Find ϕ_X for X having an (n, p) binomial distribution and calculate the first two moments of X.

Solution Decomposing X as usual by

$$X = X_1 + X_2 + \cdots + X_n$$

where the X_i are independent and all have identical two-point distributions $P_{X_i}(1) = p$ and $P_{X_i}(0) = q$, we first determine

$$\phi_{X_i}(t) = e^{it \cdot 1} p + e^{it \cdot 0} q = p e^{it} + q$$

Therefore by Proposition 6.3.4 we have

$$\phi_X(t) = (p e^{it} + q)^n$$

As for the moments of X, since X is bounded, it has finite absolute moments of all orders. Consequently by Proposition 6.3.5 we have

$$E(X^k) = \frac{1}{i^k} \phi_X^{(k)}(0)$$

Differentiating $\phi_X(t)$ with respect to t, we obtain

$$\phi_X'(t) = n(p e^{it} + q)^{n-1} p e^{it}(i)$$

Setting $t = 0$, we obtain

$$E(X) = np$$

Differentiating $\phi_X'(t)$ once more and setting $t = 0$ and $k = 2$, we obtain

$$E(X^2) = np(1 - p) + n^2 p^2$$

Example 5 Find ϕ_X for X having the standard normal distribution and whence ϕ_Y for Y having a (μ, σ) normal distribution. Prove that the sum of two independent normal variables is again normal.

Solution Using the formula

$$\phi_X(t) = E(\cos tX) + iE(\sin tX)$$

in Definition 6.3.2, we have for $f_X(x) = (1/\sqrt{2\pi})e^{-x^2/2}$

$$\phi_X(t) = \frac{1}{\sqrt{2\pi}} \int_{-\infty}^{\infty} \cos txe^{-x^2/2} \, dx + \frac{i}{\sqrt{2\pi}} \int_{-\infty}^{\infty} \sin txe^{-x^2/2} \, dx$$

The second integral vanishes since the integrand is an odd function of x. Thus we have

$$\phi_X(t) = \frac{1}{\sqrt{2\pi}} \int_{-\infty}^{\infty} \cos txe^{-x^2/2} \, dx$$

This integral is difficult to evaluate directly, so we shall instead make use of it to create an elementary differential equation for $\phi_X(t)$, which will not be difficult to solve. By differentiation with respect to t we obtain

$$\phi_X'(t) = \frac{1}{\sqrt{2\pi}} \int_{-\infty}^{\infty} (-x)\sin txe^{-x^2/2} \, dx$$

This last integral is amenable to integration by parts. Letting $u = (1/\sqrt{2\pi})\sin tx$ and $dv = -xe^{-x^2/2} \, dx$ we have

$$\phi_X'(t) = \frac{1}{\sqrt{2\pi}} \sin txe^{-x^2/2} \Big|_{-\infty}^{\infty} - \frac{1}{\sqrt{2\pi}} \int_{-\infty}^{\infty} t \cos txe^{-x^2/2} \, dx$$

$$= 0 - t\phi_X(t)$$

The differential equation $\phi_X'(t) = -t\phi_X(t)$ leads to $\phi_X'(t)/\phi_X(t) = -t$ so that $\ln \phi_X(t) = -t^2/2 + c$. The constant $c = 0$ is determined by letting $t = 0$ and noting that $\phi_X(0) = 1$. Hence, $\phi_X(t) = e^{-t^2/2}$.

If Y is (μ, σ) normal, $(Y - \mu)/\sigma = X$ is standard normal. From $Y = \sigma X + \mu$ we obtain by Proposition 6.3.4:

$$\phi_Y(t) = e^{i\mu t}e^{-\sigma^2 t^2/2} = e^{i\mu t - \sigma^2 t^2/2}$$

Likewise if W is (ν, τ) normal,

$$\phi_W(t) = e^{i\nu t - \tau^2 t^2/2}$$

so that if Y and W are independent, then

$$\phi_{Y+W}(t) = \exp\left(i\mu t - \frac{\sigma^2 t^2}{2}\right) \cdot \exp\left(ivt - \frac{\tau^2 t^2}{2}\right)$$

$$= \exp\left[i(\mu + v)t - \frac{(\sigma^2 + \tau^2)t^2}{2}\right]$$

But this is exactly the characteristic function of $(\mu + v, \sqrt{\sigma^2 + \tau^2})$ normal variable; thus according to the inversion theorem (Proposition 6.3.3) $Y + W$ must have a normal distribution with mean $\mu + v$ and variance $\sigma^2 + \tau^2$.

It is one of the neat results in probability theory that no matter how many normal variables we add the sum is always normal as long as these random variables are independent. Likewise, we point out once more that sums of independent Poisson variables are again Poisson variables (Example 3 of this section).

6.3 PROBLEMS

1. Write out the moment-generating function of a random variable having a discrete distribution at $x_1, x_2, \ldots, x_n$ with corresponding probabilities $p_1, p_2, \ldots, p_n$. Can you read off the original distribution from the moment-generating function? If X has a discrete distribution at 1, 2, 3 with probabilities p_1, p_2, p_3, and Y at $-1, -2$ with probabilities q_1, q_2, and suppose X and Y are independent, determine the probability that $X + Y = 1$ by means of moment-generating functions.

2. If the probability mass of X is found on the interval $[1, \infty]$ according to the density function $f_X(x) = 1/x^2$, show that $M_X(t)$ is not defined for $t > 0$. Hint: e^{tx}/x^2 tends to ∞ as x tends to ∞.

3. Let X have an exponential distribution $f_X(x) = \lambda e^{-\lambda x}$ for $x > 0$. Determine where $M_X(t)$ is defined.

4. The essential result in the proof of Proposition 6.3.1 is that if $M_X(t)$ is defined in some neighborhood of $t = 0$, then in this neighborhood we have

$$M_X(t) = \sum_{k=0}^{\infty} E(X^k) \frac{t^k}{k!}$$

Thus, the power series expansion of $M_X(t)$ will reveal $E(X^k)$ in the coefficients of the series. Expand the moment-generating function $e^{t^2/2}$ of the standard normal distribution in a power series of t and read off the first three moments. Hint: Let $x = t^2/2$ in the expansion $e^x = \sum_{k=0}^{\infty} x^k/k!$ and rearrange the terms.

5. Calculate the moment-generating function for a (μ, σ) normal variable directly without making use of the moment-generating function of the standard normal variable or the formula $M_{aX+b}(t) = e^{bt}M_X(at)$.

6. If the entire probability mass of X is distributed outside of the interval $(-1, 1)$ according to $f_X(x) = \frac{1}{2}(\alpha - 1)x^{-\alpha}$ where α is a positive even integer, show that $M_X(t)$ is not defined except at $t = 0$ and $M_X(0) = 1$ for $\alpha = 2, 4, 6, \ldots$ and thus it is possible for distinct distributions to have identical moment-generating functions. Hint: Cf. Problem 6.3.2.

7. Check a few of the following characteristic functions ϕ_X for X having the indicated distributions.

a. e^{iat}: one-point distribution at $x = a$.

b. $\cos at$: two-point distribution evenly at $x = \pm a$.

c. $p(1 - qe^{it})^{-1}$: p geometric distribution $P_X(k) = pq^k$ for $k = 0, 1, 2, \ldots$.

d. $e^{\mu[\exp(it) - 1]}$: μ Poisson distribution $P_X(k) = e^{-\mu}\mu^k/k!$ for $k = 0, 1, 2, \ldots$.

e. $(1 - \lambda^{-1}it)^{-1}$: λ exponential distribution $f_X(x) = \lambda e^{-\lambda x}$ for $x > 0$.

f. $(\sin at)/at$, 1 for $t = 0$: uniform distribution over $[-a, a]$.

g. $e^{-a|t|}$: Cauchy distribution $f_X(x) = a/[\pi(a^2 + x^2)]$.

h. $e^{i\mu t - \sigma^2 t^2/2}$: (μ, σ) normal distribution.

8. If X has a Cauchy distribution $f_X(x) = 1/[\pi(1 + x^2)]$ determine by means of characteristic functions the distribution of its sample average $A_n = (1/n)(X_1 + X_2 + \cdots + X_n)$. Does the law of large numbers hold for a sequence of independent random variables all identically distributed as X?

9. If X has a normal distribution, show by means of characteristic functions that its linear transform $aX + b$ with $a \neq 0$ also has a normal distribution.

10. Is the sum of two independent binomially distributed random variables binomially distributed?

11 a. If the characteristic function of X is $\phi_X(t)$, show that the characteristic function of $-X$ is

$$\phi_{-X}(t) = \check{\phi}_X(t)$$

where $\check{\phi}_X$ is the conjugate function of ϕ_X, that is, $\check{\phi}_X(t) = \overline{\phi_X(t)}$.

b. If X has a symmetric distribution in the sense that X and $-X$ have identical distributions, show that the characteristic function of X is real-valued.

c. If the characteristic function of X is real-valued, can we infer that X has a symmetric distribution?

12 a. Show that a characteristic function always satisfies

$$\phi_X(-t) = \overline{\phi_X(t)}$$

b. Show that if a characteristic function is real-valued it must be an even function, that is:

$$\phi_X(-t) = \phi_X(t)$$

Check the results of Problem 6.3.7 to confirm this.

13. Generating functions. Moment-generating functions are easier to manipulate than characteristic functions, but are less applicable in the sense that some random variables do not have moment-generating functions. Going in the direction of easier manipulability but of less applicability, we have the so-called generating functions, which are even easier to manipulate than the moment-generating functions but are less applicable. Generating functions are defined only for the random variables which take only the nonnegative integer values. For a random variable X with $P(X=j)=a_j$ for $j=0, 1, 2, 3, \ldots$, we define its generating function by

$$G_X(s) = \sum_{j=0}^{\infty} a_j s^j \qquad \text{for} \quad |s| \leq 1$$

a. Show that $\sum_{j=0}^{\infty} a_j s^j$ converges absolutely for $|s| \leq 1$ by comparison with a geometric series.

b. Show that if the expected value of X is finite, then

$$E(X) = G_X'(1)$$

where $G_X'(1)$ is the derivative of $G_X(s)$ with respect to s evaluated at $s=1$.

c. Show that if the variance of X exists, we have

$$\text{var}(X) = G_X''(1) + G_X'(1) - G_X'^2(1)$$

Hint: $G_X''(1) = E(X^2) - E(X)$.

d. Show that if X and Y have an (n, p) binomial distribution and a μ Poisson distribution, respectively, then

$$G_X(s) = (q + ps)^n$$
$$G_Y(s) = e^{-\mu + \mu s}$$

e. Use generating functions to calculate the mean and the variance for the (n, p) binomial distribution and μ Poisson distribution. Would you have obtained these results just as easily if you used the moment-generating functions instead?

f. If X and Y are both nonnegative integer-valued random variables with generating functions G_X and G_Y, and are independent, can you obtain the generating function of $X + Y$ from G_X and G_Y? (Cf. Problem 3.6.4.)

6.4 CONVERGENCE IN DISTRIBUTION

Up to now two given distributions are either identical or different, and we have not bothered to examine the possible proximity that might exist between these distributions. Yet, given a sequence of distributions $P_1, P_2, P_3, \ldots$ all differing from a fixed distribution P, P_n may nevertheless become less and less distinguishable from P as n gets larger in the sense that $P_n[a, b]$ tends to $P[a, b]$ for each interval $[a, b]$. For example if P_n is a uniform distribution over the interval $[-1 - 1/n, 1 + 1/n]$ and P a uniform distribution over the interval $[-1, 1]$, it is not difficult to see that P_n approaches P in the sense described above. Accordingly we consider the following somewhat tentative definition.

6.4.1 DEFINITION

A sequence of probability distributions P_n is said to converge (in a strict sense) to a probability distribution P if

$$\lim_{n \to \infty} P_n[a, b] = P[a, b]$$

for each finite closed interval $[a, b]$ with $a \leq b$ in R.

It can be shown that the convergence of P_n to P on every closed finite interval as postulated above implies that P_n converges to P on every finite interval, not necessarily closed (see Problem 6.4.1). Definition 6.4.1 however is unnecessarily strict, for it will not permit us to view P_n with $P_n(1/n) = 1$ as converging to P with $P(0) = 1$ since, for example, $P_n[-1, 0] = 0$ for all n and thus will not approach $P[-1, 0] = 1$. Accordingly we consider the following less strict and hence more generally acceptable definition.

6.4.2 DEFINITION

A sequence of probability distributions P_n is said to converge to a probability distribution P if

$$\lim_{n \to \infty} P_n[a, b] = P[a, b]$$

for every finite closed interval $[a, b]$ where the end points $a \leq b$ are in some "thick" subset D of R. A subset D of R is said to be *thick* if it consits of all but countably many real numbers.

Thus we are only requiring P_n to approach P on many (but not all) finite closed intervals. With this definition we can now say that P_n with $P_n(1/n) = 1$ converges to P with $P(0) = 1$ since we need only let D be the set of all real numbers except 0. Note also that convergence of P_n to P on every finite closed interval with end points in D will automatically guarantee the convergence of P_n to P on every finite interval with end points in D (see Problem 6.4.1 again).

The convergence concept of probability distributions naturally has some bearing on random variables.

6.4.3 DEFINITION

Given a sequence of random variables X_n and a random variable X all defined on the same probability space, let P_n and P denote the probability distribution of X_n and X, respectively. Then X_n is said to *converge in distribution* to X (or to P if you like) whenever P_n converges to P.

Convergence in distribution of X_n to X is just that and nothing more; random variables X_n as functions on the probability space need not converge to X in any other way. Consider for example a sequence of indicator functions $X_n = I_A$ for all odd n and $X_n = I_B$ for all even n where A and B partition the sample space S of the probability space $(S, \mathscr{S}, P)$. If $P(A) = P(B) = \frac{1}{2}$, then X_n converges in distribution to I_A as well as I_B since P_n are all identical two-point distributions;

yet, for any fixed point ω in S, $X_n(\omega)$ oscillates forever between 0 and 1, preventing X_n from converging pointwise to any X.

Convergence in distribution of X_n to X does however, imply the pointwise convergence of the distribution functions F_n to F (well, not quite but almost, as we shall soon see). The converse of this, to be formulated more precisely later, then enables us to determine the convergence in distribution of X_n to X from examining the pointwise convergence of F_n to F. We must now proceed with caution and patience.

6.4.1 PROPOSITION

If X_n converges in distribution to X, then

$$\lim_{n \to \infty} F_n(x) = F(x)$$

for all x in some thick (see Definition 6.4.2) subset D of R, in particular for all x in R if X_n converges in distribution to X in the strict sense of Definition 6.4.1.

PROOF (Optional)

For any x in D as postulated in Definition 6.4.2 we must show

$$\lim_{n \to \infty} P_n(-\infty, x] = P(-\infty, x]$$

Now, as long as a is also in D, we have

$$\lim_{n \to \infty} P_n(a, x] = P(a, x]$$

but since $P_n(-\infty, x] \geqq P_n(a, x]$ for every n, we see

$$\liminf_{n \to \infty} P_n(-\infty, x] \geqq P(a, x]$$

Since this is true for any a in D, by letting a move to $-\infty$ in D we obtain

$$\liminf_{n \to \infty} P_n(-\infty, x] \geqq P(-\infty, x]. \tag{1}$$

Similarly we can show (do as an exercise)

$$\liminf_{n \to \infty} P_n(x, \infty) \geqq P(x, \infty) \tag{2}$$

but $P_n(x, \infty) = 1 - P_n(-\infty, x]$ and $P(x, \infty) = 1 - P(-\infty, x]$, consequently we can rewrite (2) as

$$\liminf_{n \to \infty} [1 - P_n(-\infty, x]] \geqq 1 - P(-\infty, x]$$

or

$$1 - \limsup_{n \to \infty} P_n(-\infty, x] \geqq 1 - P(-\infty, x]$$

hence

$$\limsup_{n \to \infty} P_n(-\infty, x] \leq P(-\infty, x]$$

Combining this last inequality with (1) we obtain

$$\lim_{n \to \infty} P_n(-\infty, x] = P(-\infty, x]$$

6.4.1 COROLLARY

If X_n converges in distribution to X, then

$$\lim_{n \to \infty} F_n(x) = F(x)$$

for all x at which F is continuous.

PROOF (Optional)

At a continuity point x of F take a δ-neighborhood so small that for any y in this neighborhood we have

$$|F(y) - F(x)| < \varepsilon$$

for some arbitrarily given $\varepsilon > 0$. From this δ-neighborhood choose in particular two points $y' < x < y''$ belonging to the thick set D of Proposition 6.4.1. Then from

$$F_n(y') \leq F_n(x) \leq F_n(y'')$$

it follows that

$$F(y') = \lim_{n \to \infty} F_n(y') \leq \liminf_{n \to \infty} F_n(x) \leq \limsup_{n \to \infty} F_n(x) \leq \lim_{n \to \infty} F_n(y'') = F(y'')$$

But $F(y')$ and $F(y'')$ are both within ε distance from $F(x)$, therefore the same must hold for $\liminf F_n(x)$ and $\limsup F_n(x)$. Now since ε is arbitrarily small, this means

$$\liminf_{n \to \infty} F_n(x) = \limsup_{n \to \infty} F_n(x) = F(x)$$

and the corollary follows.

We now state a useful converse of Proposition 6.4.1 as follows.

6.4.2 PROPOSITION

Given X_n and X, if

$$\lim_{n \to \infty} F_n(x) = F(x)$$

for all x in some thick subset D of R, then X_n converges in distribution to X.

PROOF (Optional)

Remove from the set D all points of discontinuity of F. Call this set D', then D' is still a thick set since F being a monotone function can have at most countably many points of discontinuity. For any $a \leq b$ in D' we shall show

$$P_n[a, b] \to P[a, b] \tag{1}$$

since this will establish the convergence in distribution of X_n to X.

First we show easily that

$$P_n(a, b] \to P(a, b] \tag{2}$$

This follows from $P_n(a, b] = F_n(b) - F_n(a)$, $P(a, b] = F(b) - F(a)$, and $F_n(b) \to F(b)$, $F_n(a) \to F(a)$ by assumption.

Now consider

$$P_n(y'', b] \leq P_n[a, b] \leq P_n(y', b]$$

where $y' < a < y'' < b$ are all points in D'. By (2) we have

$$P(y'', b] \leq \liminf_{n \to \infty} P_n[a, b] \leq \limsup_{n \to \infty} P_n[a, b] \leq P(y', b]$$

but we could have chosen y'' and y' sufficiently close to a so that

$$P(y', b] - P[a, b] < \varepsilon$$

$$P[a, b] - P(y'', b] < \varepsilon$$

for any arbitrarily small $\varepsilon > 0$. Note that the last inequality does not necessarily follow unless $P(a) = 0$ as guaranteed by the continuity of F at a (cf. Proposition 3.2.4). Since both $P(y', b]$ and $P(y'', b]$ can be made arbitrarily close to $P[a, b]$, this will make $\liminf P_n[a, b] = \limsup P_n[a, b]$, from which (1) follows.

6.4.2 COROLLARY

Given X_n and X, if

$$\lim_{n \to \infty} F_n(x) = F(x)$$

for all x at which F is continuous, then X_n converges in distribution to X.

PROOF (Optional)

Let D be the set of points at which F is continuous. Since F being monotone can have at most countably many points of discontinuity, D is a thick set and the corollary follows by the preceding proposition.

We have just seen how convergence in distribution of X_n to X is reflected by a nearly perfect pointwise convergence of F_n to F. We can state specifically that given a sequence of random variables X_n, if their distribution functions F_n tend to some function F for all x in some thick subset of R (or at all x where F is continuous) and if F happens (see Problem 6.4.5) to be indeed a distribution function (see Section 3.2 for the three conditions F must satisfy), then X_n converges in distribution to F (or shall we say to some X having F as its distribution function). Now characteristic functions provide another means of describing probability distributions for random variables beside distribution functions. We can therefore investigate just how convergence in distribution of X_n to X is reflected by the corresponding characteristic functions ϕ_n and ϕ. We shall state the known results in this area without proof (see Minibibliography: Chung, Gnedenko). We merely point out that these results appear to be quite plausible by comparison with the corresponding results involving the distribution functions (Propositions 6.4.1 and 6.4.2). The proofs of these results actually make use of the results already established for distribution functions.

6.4.3 PROPOSITION

If X_n converges in distribution to X, then

$$\lim_{n \to \infty} \phi_n(x) = \phi(x)$$

for each x in R. In fact ϕ_n converges to ϕ uniformly in every finite interval.

Note how neat this result is compared to that of Proposition 6.4.1.

As a useful converse to Proposition 6.4.3 we have the following proposition.

6.4.4 PROPOSITION

Given X_n and X, if

$$\lim_{n \to \infty} \phi_n(x) = \phi(x)$$

for every x in R, then X_n converges in distribution to X.

Again note the analogy of this proposition to Proposition 6.4.2. Proposition 6.4.4 as stated here is not very useful, however, since it presupposes the knowledge of some X while in practice we may have only X_n at hand to examine. This means that in general after determining $\lim_{n \to \infty} \phi_n(x)$ we must ascertain whether this limit is indeed a characteristic function of some probability distribution. In other words, we must characterize characteristic functions in much the same way as we characterized distribution functions by listing the axiomatic conditions these functions must satisfy. Such conditions have been found by Bochner and Herglotz (see Minibibliography: Chung, p. 164). According to them a complex valued function ϕ of a real variable x is a characteristic function (of some probability distribution) if and only if

(1) ϕ is " positive definite " (Chung, p. 165);
(2) ϕ is continuous at $x = 0$ with $\phi(0) = 1$.

The condition (1) is hard to verify; fortunately it turns out to be redundant for functions that are actually limits of characteristic functions. In fact we have the following powerful result due to Lévy and Cramér.

6.4.1 THEOREM (Lévy-Cramér)

Given a sequence of random variables X_n, if their characteristic functions ϕ_n converge to some function ϕ everywhere in R and ϕ happens to be continuous at $x = 0$, then the probability distributions P_n of X_n converge to a probability distribution P, whose characteristic function is ϕ.

Let us apply this theorem to the following interesting situation. We know that if X_n is (n, p) binomially distributed for $n = 1, 2, 3, \ldots$ where p is fixed, the probability mass of X_n will "escape to infinity" as n tends to ∞ since the mean value np of X_n will approach ∞ while the standard deviation $\sqrt{npq}$ does not "spread out" quite fast enough to allow enough mass to "flow back" toward the origin. Therefore, in order to arrest the probability mass, let us suppose that X_n is (n, p_n) binomially distributed for $n = 1, 2, 3, \ldots$ with p_n shrinking to 0 in such a way that

$$np_n = \mu$$

for all n where μ is some fixed constant. Now the probability mass of X_n cannot quite escape to infinity as n approaches ∞ since the mean value of X_n is μ for all n, and we just might be able to determine the "ultimate" distribution of X_n. We settle this by the following example.

Example 1 Let X_n be such that its distribution P_n is $(n, \mu/n)$ binomial for all n. Determine the distribution P, if any, to which P_n converges.

Solution We know that if X_n is (n, p) binomially distributed, its characteristic function is given by

$$\phi_n(t) = (pe^{it} + q)^n$$
$$= [1 + p(e^{it} - 1)]^n$$

Now since X_n is $(n, \mu/n)$ binomially distributed, we let $p = \mu/n$ to obtain

$$\phi_n(t) = \left[1 + \frac{\mu(e^{it} - 1)}{n}\right]^n$$

Consequently we have (remembering from calculus that $\lim_{n \to \infty}(1 + a/n) = e^a$)

$$\lim_{n \to \infty} \phi_n(t) = e^{\mu(e^{it} - 1)}$$

which certainly is continuous at $t = 0$, hence P_n must converge to some distribution P by Theorem 6.4.1. As a matter of fact the limiting function is just the characteristic function of μ Poisson distribution [see Problem 6.3.7(d)] so that Proposition 6.4.4 suffices in this case as in many important special cases.

Example 1 amounts to a nonelementary proof of the well-known Poisson limit theorem (see Section 2.6 also).

Example 2 (Poisson Approximation) If on the average 1 out of 1000 transistors manufactured is found to be defective, use the Poisson limit theorem to find approximately the probability that in a case of 5000 transistors more than 5 are defective.

Solution Interpreting " 1 defective out of a 1000 " as meaning that an arbitrarily picked transistor has probability $p = 0.001$ of being defective, we identify the number X of defectives in 5000 with the number of heads in 5000 tosses of a coin with probability $p = 0.001$ of falling heads. Thus X has an (n, p) binomial distribution with $n = 5000$ and $p = 0.001$. In view of Poisson limit theorem we see that X has approximately μ Poisson distribution with $\mu = np = 5$. Consequently,

$$P(X > 5) \doteq 1 - \sum_{k=0}^{5} e^{-5} \frac{5^k}{k!} \doteq 0.384$$

6.4 PROBLEMS

1. Show that if $P_n[a, b]$ tends to $P[a, b]$ as n tends to ∞ for every finite closed interval $[a, b]$ with $a \leq b$ in R then $P_n(I)$ tends to $P(I)$ for any finite interval I. Hint: $(a, b] = [c, b] - [c, a]$ for some $c < a$.

2. Show that if P_n is a uniform distribution over the interval $[-1/n, 1/n]$ and P is a one-point distribution at the origin, $P(0) = 1$, then P_n converges to P. Does P_n converge to P in the strict sense of Definition 6.4.1. Why not?

3. Let P_n be a one-point distribution, $P_n(n) = 1$. Does P_n converge to any distribution at all?

4. Let P_n be an (n, p) binomial distribution with p fixed. Does P_n converge to any distribution at all?

5. Let X_n have a one-point distribution at n, that is, $P_n(n) = 1$ for $n = 1, 2, 3, \ldots$. Determine $\lim_{n \to \infty} F_n(x)$. Is the limiting function a distribution function? Why not?

6. Let P_n be a uniform distribution over the interval $[-1/n, 1/n]$. Determine by means of characteristic functions the distribution P to which P_n converges. Do the same for P_n, a uniform distribution over the interval $[-1 - 1/n, 1 + 1/n]$. Hint: See Problem 6.3.7(a) and (f).

6.5 CENTRAL LIMIT THEOREMS

If a sequence of probability distributions $P_1, P_2, P_3, \ldots$ is such that $\lim_{n \to \infty} P_n(I) = 0$ for every finite interval I, we say that the probability mass escapes to ∞. On the other hand if P_n converges to a one-point distribution P with $P(a) = 1$ for some real number a, we might say that the probability mass escapes into a point. In either case it is impossible to discern the pattern of distribution (such as Poisson or normal), if any, that P_n may tend to have as n approaches ∞, for it is actually quite possible that while P_n converges to a one-point distribution or escapes to ∞, P_n resembles more and more a Poisson distribution or a normal distribution.

Now given a sequence of random variables $S_1, S_2, S_3, \ldots$, the simplest way to prevent their probability distributions $P_1, P_2, P_3, \ldots$ from escaping to ∞ or into a point is to standardize $S_1, S_2, S_3, \ldots$ to $Z_1, Z_2, Z_3, \ldots$ where

$$Z_n = \frac{S_n - \mu_n}{\sigma_n}$$

with μ_n and σ_n being the mean value and the standard deviation of S_n. Then, if Z_n converges in distribution to say a normal distribution, we can say that S_n is just about normally distributed for large n. The well-known assertion of central importance in probability theory and statistical theory, generally referred to as the central limit theorem, describes just such a phenomenon.

6.5.1 THEOREM

Given a sequence of independent random variables $X_1, X_2, X_3, \ldots$ all of which are identically distributed as a certain random variable X with mean μ and standard deviation σ, if $S_1, S_2, S_3, \ldots$ are the sequence of the partial sums

$$S_n = X_1 + X_2 + \cdots + X_n$$

then the standardized sequence

$$Z_n = \frac{S_n - n\mu}{\sqrt{n}\sigma}$$

converges in distribution to the standard normal variable.

Before we go into the proof of this particular central limit theorem, we note that although the assertion of the theorem is about Z_n, what is really important for us is its implication about S_n, namely that the sum S_n of a large number of identically distributed independent random variables having finite means μ and standard deviations σ is approximately $(n\mu, \sqrt{n}\sigma)$ normally distributed. For example, the weight of an individual from a certain population is the sum of the weight of various biological components forming the individual, hence it should be approximately normally distributed. Or the error in measurement of a certain fixed physical quantity is the sum of many small independent error factors, hence it too must be approximately normally distributed. These and many other examples however do also point out the need to generalize Theorem 6.5.1 since the summand random variables are often independent without being quite identically distributed, or are not independent at all.

PROOF

In view of Proposition 5.4.4 we need only show that the characteristic function $\phi_n(t)$ of Z_n approaches $e^{-t^2/2}$, the characteristic function of standard normal distribution. We calculate $\phi_n(t)$ as follows. First rewrite

$$Z_n = \frac{(X_1 - \mu) + (X_2 - \mu) + \cdots + (X_n - \mu)}{\sqrt{n}\sigma}$$

so that we see that Z_n is essentially the sum of n independent random variables all having the same distribution as $X - \mu$. Let $\phi(t)$ be the characteristic function for $X - \mu$, then the characteristic function for $\sqrt{n}\sigma Z_n$ is simply $[\phi(t)]^n$, and the characteristic function for Z_n itself is

$$\phi_n(t) = [\phi(t/\sqrt{n}\sigma)]^n$$

by Proposition 6.3.4. To determine $\lim_{n \to \infty} \phi_n(t)$ we first work out a suitable Taylor expression for $\phi(t/\sqrt{n}\sigma)$. We recall from calculus that if a function $\phi(t)$ has a kth derivative in a neighborhood of $t = 0$, then $\phi(t)$ can be expanded into

$$\phi(t) = \sum_{j=0}^{k} \frac{\phi^{(j)}(0)}{j!} t^j + o(|t|^k)$$

where $o(|t|^k)$ denotes a quantity that goes to 0 faster than $|t|^k$, that is, $\lim_{t \to \infty} o(|t|^k)/|t|^k = 0$. To apply this result to our $\phi(t)$ and thence to $\phi(t/\sqrt{n}\sigma)$, note that since $X - \mu$ has a finite second-order absolute moment, which is

$$E(|X - \mu|^2) = \sigma^2$$

by Proposition 6.3.5 its characteristic function $\phi(t)$ has a second derivative, certainly in a neighborhood of $t = 0$, therefore we have

$$\phi\left(\frac{t}{\sqrt{n}\sigma}\right) = 1 + \frac{\phi'(0)}{1!} \frac{t}{\sqrt{n}\sigma} + \frac{\phi''(0)}{2!}\left(\frac{t}{\sqrt{n}\sigma}\right)^2 + o\left(\frac{t^2}{n\sigma^2}\right)$$

But $\phi'(0) = i \cdot E(X - \mu) = 0$ and $\phi''(0) = i^2 \cdot E(X - \mu)^2 = -\sigma^2$ (by Proposition 6.3.5), hence we have

$$\phi\left(\frac{t}{\sqrt{n}\sigma}\right) = 1 - \frac{t^2}{2n} + o\left(\frac{t^2}{n\sigma^2}\right)$$

Thus,

$$\lim_{n \to \infty} \phi_n(t) = \lim_{n \to \infty} \left[1 + \frac{1}{n}\left(-\frac{t^2}{2}\right) + o\left(\frac{t^2}{n\sigma^2}\right)\right]^n$$

$$= e^{-t^2/2}$$

The last equality can be justified by the following result from calculus: if $\lim_{n \to \infty} a_n = a$, then

$$\lim_{n \to \infty} \left(1 + \frac{1}{n} a_n\right)^n = e^a$$

Example 1 A certain "random machine" responds to each operation with an unpredictable real number X. After a long series of n operations the average A_n of these n real numbers $X_1, X_2, \ldots X_n$ is calculated

$$A_n = \frac{X_1 + X_2 + \cdots + X_n}{n}$$

Show that it is reasonable to assume that A_n is approximately normal regardless of what distribution X has, as long as X has a finite mean and a finite standard deviation (which is usually assumed).

Solution According to Theorem 6.5.1, $X_1 + X_2 + \cdots + X_n = S_n$ is approximately normal. Using characteristic functions (Problem 6.3.7), it is easy to show that if a random variable Z is (μ, σ) normally distributed, the random variable $aZ + b$ is $(a\mu + b, a\sigma)$ normally distributed. Thus, since $A_n = (1/n)S_n$ is a linear transform of S_n and S_n is approximately normal, it is reasonable to assume that A_n is also approximately normal.

Alternatively, we observe that the standardization of A_n is no different from the standardization of S_n (Problem 6.5.1), and hence by Theorem 6.5.1 A_n is approximately normal for large values of n just as S_n is.

As a well-known special case of Theorem 6.5.1, we let X have a two-point distribution, $P_X(1) = p$ and $P_X(0) = q$ with $p + q = 1$; in other words X is the random variable representing a coin with probability p of falling heads so that $S_n = X_1 + X_2 + \cdots + X_n$ is the random variable counting the number of heads in n tosses of such a coin. Since S_n has an (n, p) binomial distribution, Theorem 6.5.1 tells us that for large values of n (n, p) binomial distributions are approximately normal. This is the well-known Demoivre-Laplace limit theorem, which we state formally below.

6.5.1 COROLLARY (Demoivre-Laplace)
If S_n has an (n, p) binomial distribution with p fixed for all n

$$P(S_n = r) = C_r^n p^r(1 - p)^{n-r} \qquad \text{for} \quad r = 0, 1, 2, \ldots, n$$

then $Z_n = (S_n - np)/\sqrt{npq}$ converges in distribution to the standard normal distribution; specifically we have

$$\lim_{n \to \infty} p\left(a \leq \frac{S_n - np}{\sqrt{npq}} \leq b\right) = \frac{1}{\sqrt{2\pi}} \int_a^b e^{-x^2/2} \, dx \qquad (1)$$

for all real $a \leq b$.

PROOF

In order to apply Theorem 6.5.1 we need only indicate that S_n can be regarded as a sum of independent random variables all having identical distributions, which we already did before stating the corollary, and that np and $\sqrt{npq}$ are the mean and the standard deviation of S_n so that Z_n is indeed the standardization of S_n. We point out also that the limit statement (1) holds for all real $a \leq b$ because the normal distribution function is continuous everywhere (see Corollary 6.4.1).

The limit statement (1) is of tremendous help in approximating probability values pertaining to a binomial variable. Let us elaborate on this. Suppose that a random variable S has an (n, p) binomial distribution and that we want to determine as accurately as practicable the probability value $P(r_1 \leq S \leq r_2)$ where r_1 and r_2 are nonnegative integers. The exact calculation

$$P(r_1 \leq S \leq r_2) = \sum_{r=r_1}^{r_2} C_r^n p^r (1 - p)^{n-r}$$

may be too cumbersome, especially for large values of n and $r_2 - r_1$, so we turn to approximation by Demoivre-Laplace limit theorem. For reasons that shortly will become clear we deliberately equate

$$P(r_1 \leq S \leq r_2) = P(r_1 - \tfrac{1}{2} \leq S \leq r_2 + \tfrac{1}{2}) \tag{2}$$

noting that probability mass is found only at integer points anyway. Then we make a translation

$$P(r_1 - \tfrac{1}{2} \leq S \leq r_2 + \tfrac{1}{2}) = P\left(\frac{r_1 - \tfrac{1}{2} - np}{\sqrt{npq}} \leq \frac{S - np}{\sqrt{npq}} \leq \frac{r_2 + \tfrac{1}{2} - np}{\sqrt{npq}}\right) \tag{3}$$

which by (1) is approximately equal to

$$\Phi\left(\frac{r_2 + \tfrac{1}{2} - np}{\sqrt{npq}}\right) - \Phi\left(\frac{r_1 - \tfrac{1}{2} - np}{\sqrt{npq}}\right) \tag{4}$$

where $\Phi(x)$ is the standard normal distribution function whose values can be found in tables.

It is convenient to see graphically what is happening when a binomial distribution is being approximated by a normal distribution. By the Demoivre-Laplace limit theorem, if S is (n, p) binomial, $(S - np)/\sqrt{npq}$ is nearly $(0, 1)$ normal. We express this by writing

$$\frac{S - np}{\sqrt{npq}} \doteq Z$$

where Z is a $(0, 1)$ normal random variable. From this we see

$$S \doteq \sqrt{npq}\, Z + np$$

that is, S is approximately $(np, \sqrt{npq})$ normal. Denoting such a normal variable by Z', we can compare the distributions of S and Z' by looking at their distribution functions (see Fig. 6.5.1).

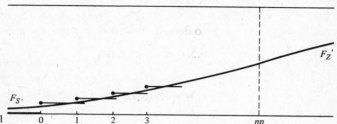

Figure 6.5.1

The intertwining nature of the graphs of F_S and $F_{Z'}$ illustrates why $P(2 \leq S \leq 3)$ for instance is better approximated by $F_{Z'}(3 + \frac{1}{2}) - F_{Z'}(2 - \frac{1}{2})$ than by $F_{Z'}(3) - F_{Z'}(2)$. Now, of course

$$F_{Z'}(r_2 + \tfrac{1}{2}) - F_{Z'}(r_1 - \tfrac{1}{2}) = \Phi\left(\frac{r_2 + \tfrac{1}{2} - np}{\sqrt{npq}}\right) - \Phi\left(\frac{r_1 - \tfrac{1}{2} - np}{\sqrt{npq}}\right)$$

Example 2 If a coin with probability $p = \frac{1}{5}$ of falling heads is tossed 10,000 times, estimate by normal approximation the probability that the number of heads will be between 1900 and 2100.

Solution The number S of heads in 10,000 tosses is $(10,000, \frac{1}{5})$ binomial and approximately $(2000, 40)$ normal since

$$np = 10,000 \times \tfrac{1}{5} = 2000$$
$$\sqrt{npq} = \sqrt{10,000 \cdot \tfrac{1}{5} \cdot \tfrac{4}{5}} = 40$$

Hence,

$$P(1900 \leq S \leq 2100) \doteq \Phi\left(\frac{2100 - 2000}{40}\right) - \Phi\left(\frac{1900 - 2000}{40}\right)$$
$$= \Phi(2.5) - \Phi(-2.5)$$
$$\doteq 0.988$$

We see that the probability is quite close to 1.

The result of this example is not really surprising in view of the weak law of large numbers since

$$P(1900 \leq S \leq 2100) = P\left(\frac{1900}{10,000} \leq \frac{S}{10,000} \leq \frac{2100}{10,000}\right)$$
$$= P(\tfrac{1}{5} - \tfrac{1}{100} \leq A \leq \tfrac{1}{5} + \tfrac{1}{100})$$
$$\doteq 1$$

where A is what we once called the sample average (see Section 6.1), and found to cluster around the mean value, $\frac{1}{5}$ in this case.

We recall and note that the law of large numbers says something about the sample average A_n while the central limit theorem says something about sample sum S_n, but then A_n and S_n are related by

$$A_n = \frac{S_n}{n}$$

and therefore we ought to be able to deduce the law of large numbers (specifically the weak form) by investigating the limit of the characteristic functions $\psi_n(t)$ of A_n. Now in the proof of Theorem 6.5.1 we have seen the limit of the characteristic functions $\phi_n(t)$ of $Z_n = (S_n - n\mu)/\sqrt{n}\sigma$ to be $e^{-t^2/2}$. But

$$A_n = \frac{1}{n} S_n = \frac{1}{n} (\sqrt{n}\sigma Z_n + n\mu)$$

$$= \frac{\sigma}{\sqrt{n}} Z_n + \mu$$

hence

$$\psi_n(t) = e^{i\mu t}\phi_n\left(\frac{\sigma t}{\sqrt{n}}\right)$$

Thus

$$\lim_{n\to\infty} \psi_n(t) = e^{i\mu t} \lim_{n\to\infty} \phi_n\left(\frac{\sigma t}{\sqrt{n}}\right)$$

$$= e^{i\mu t} e^{-0^2/2}$$

The last step is justified by the following general consideration: If continuous functions $\phi_n(t)$ converge uniformly to $\phi_0(t)$ and t_n converge to t_0, then $\phi_n(t_n)$ converge to $\phi_0(t_0)$. Recalling that $e^{i\mu t}$ is the characteristic function of a one-point distribution (Problem 5.3.6a), we obtain the following version of the weak law of large numbers.

6.5.1 PROPOSITION (A Weak Law of Large Numbers)

Given a sequence of independent random variables $X_1, X_2, \ldots$ all identically distributed as a random variable X with mean value μ and standard deviation σ, if

$$A_n = \frac{X_1 + X_2 + \cdots + X_n}{n}$$

then $A_1, A_2, \ldots$ converge in distribution to the one-point distribution at μ.

Thus the probability mass of A_n escapes to the point μ. Note, however, like S_n, which escapes to ∞, A_n for large n is approximately $(\mu, \sigma/\sqrt{n})$ normally distributed. The fact that the sample average A_n for large n is approximately normally distributed is of great importance to statisticians.

The rest of this section consists of optional material dealing with some possible generalizations of Theorem 6.5.1.

The core assertion of Theorem 6.5.1 is that the sequence of partial sums S_n of a given sequence of random variables X_n tend toward having a normal distribution in the sense that their standardizations Z_n converge in distribution to the standard normal distribution. We state this formally as follows.

6.5.1 DEFINITION

A sequence of random variables $X_1, X_2, X_3, \ldots$ is said to *obey the central limit law* if the standardizations of the partial sums of the sequence converge in distribution to the standard normal distribution, where by partial sums $S_1, S_2, S_3, \ldots$ we mean

$$S_n = X_1 + X_2 + \cdots + X_n$$

and by standardizations $Z_1, Z_2, Z_3, \ldots$ we mean

$$Z_n = \frac{S_n - \mu(S_n)}{\sigma(S_n)}$$

From the standpoint of this definition Theorem 6.5.1 simply says that $X_1, X_2, X_3, \ldots$ obey the central limit law if
(1) $X_1, X_2, X_3, \ldots$ are independent
(2) $X_1, X_2, X_3, \ldots$ are all identically distributed with finite means and variances

Now granted that (1) and (2) constitute a *sufficient* set of conditions for the central limit law to hold, we may nevertheless question the *necessity* of these two conditions. The removal of condition (2) is especially desirable since one often encounters sums of a large number of independent random variables that are not necessarily identically distributed. For example in the theory of error one may regard the error in measurement of some physical quantity as the sum of many kinds of small independent errors whose individual distributions may be quite diverse and difficult to determine. If, however, the central limit law can be shown to apply, then for all practical purposes one may regard the total error to be normally distributed. This turned out to be indeed the case in view of results due to Liapounov (1901) and Lindeberg (1922). Lindeberg's condition (see Theorem 6.5.2) is about the most general condition known that guarantees the central limit law for sequences of independent random variables. Among all the central limit theorems asserting the validity of central limit law under various assumptions, Corollary 6.5.1 is historically the first though understandably the least general. It is due to Demoivre (1732) for $p = \frac{1}{2}$ and to Laplace (1812) for $0 < p < 1$.

6.5.2 THEOREM (Lindeberg)

A sequence of independent random variables $X_1, X_2, X_3, \ldots$ obeys the central limit law if for any $\varepsilon > 0$

$$\lim_{n \to \infty} \frac{1}{B_n^2} \sum_{k=1}^{n} \int_{|x - \mu_k| < \epsilon B_n} (x - \mu_k)^2 P_k(dx) = 1$$

where $B_n^2 = \text{var}(X_1 + X_2 + \cdots + X_n)$, $\mu_k = \mu(X_k)$, and P_k is the distribution of X_k.

Note that for each integer k the integral above can be regarded as a "fragment" of the variance of X_k since

$$\text{var}(X_k) = \int_R (x - \mu_k)^2 P_k(dx)$$

To see what Lindeberg's condition really means let us consider in general the following what we might call the "ε-fragment of variance" of a random variable X:

$$\text{var}_{(\epsilon)}(X) = \int_{|x - \mu_X| \leqslant \epsilon} (x - \mu_X)^2 P_X(dx)$$

and rewrite the Lindeberg expression in Theorem 6.5.2 as follows

$$\sum_{k=1}^{n} \int_{|(x - \mu_k)/B_n| < \epsilon} [(x - \mu_k)/B_n]^2 P_k(dx) = \sum_{k=1}^{n} \int_{|y| < \epsilon} y^2 P_{n,k}(dy)$$

$$= \sum_{k=1}^{n} \text{var}_{(\epsilon)}(Y_{n,k})$$

where $Y_{n,k}$ denotes the random variable $(X_k - \mu_k)/B_n$ and $P_{n,k}$ its distribution induced from P_k via the mapping $y = (x - \mu_k)/B_n$ (cf. Section 4.2). We note that for each fixed n there are n such random variables $Y_{n,1}, Y_{n,2}, \ldots, Y_{n,n}$ and that their sum $Y_{n,1} + Y_{n,2} + \cdots + Y_{n,n}$ is none other than the standardization Z_n of $S_n = X_1 + X_2 + \cdots + X_n$, so that

$$\sum_{k=1}^{n} \text{var}(Y_{n,k}) = \text{var}\left(\sum_{k=1}^{n} Y_{n,k}\right) = 1$$

It is against this background that we now consider the Lindeberg condition

$$\sum_{k=1}^{n} \text{var}_{(\epsilon)}(Y_{n,k}) \to 1$$

as $n \to \infty$ regardless of how small ε is. Now admittedly

$$\sum_{k=1}^{n} \mathrm{var}_{(\epsilon)}(Y_{n,k}) \leqq \sum_{k=1}^{n} \mathrm{var}(Y_{n,k})$$

and the left-hand side of the inequality can be much smaller than the right-hand side for each fixed n, especially when ε is a very small quantity. Lindeberg's condition demands however that as n approaches ∞ their discrepancy become arbitrarily small. Let us investigate how such a demand can possibly be satisfied. Quite obviously, in order for the Lindeberg "phenomenon" to take place $\mathrm{var}_{(\epsilon)}(Y_{n,k})$ must capture most of $\mathrm{var}(Y_{n,k})$ for all $k = 1, 2, 3, \ldots, n$ for large values of n (and this regardless of how small ε is). This however will be quite impossible unless the variances of $Y_{n,k}$ (for $k = 1, 2, \ldots, n$) all approach 0 (with some uniformity) as n approaches ∞. In other words we must insist that

$$\max_{k \leqq n} \mathrm{var}(Y_{n,k}) \to 0$$

as $n \to \infty$. Now since

$$\mathrm{var}(Y_{n,k}) = \mathrm{var}\left(\frac{X_k - \mu_k}{B_n}\right) = \frac{\sigma_k^2}{B_n^2}$$

we can formally state the following proposition.

6.5.2 PROPOSITION
In order for Lindeberg's condition to hold we must (at least) have

$$\lim_{n \to \infty} \max_{k \leqq n} \sigma_k^2 / B_n^2 = 0$$

Let us call this weaker condition Feller's condition. Feller's condition is quite easy to interpret. It merely says that in order for a sequence of independent random variables $X_1, X_2, X_3, \ldots$ to satisfy Lindeberg's condition the variances of summand random variables $X_1, X_2, X_3, \ldots, X_n$ must be uniformly small relative to the variance of the partial sum S_n and that this relative smallness be intensified as n approaches ∞. Now Lindeberg's condition is a sufficient condition for the central limit law whereas Feller's condition is a necessary condition for Lindeberg's condition, and this makes Feller's condition neither necessary nor sufficient for central limit law. Nevertheless Feller's condition acquires its importance in view of the following dichotomy:

(a) Among all sequences of independent random variables satisfying Feller's condition those which obey the central limit law are precisely those which satisfy Lindeberg's condition.
(b) Among all sequences of independent random variables not satisfying Feller's condition those which obey the central limit law are precisely those whose summand random variables are almost all normally distributed.

Statement (a) is due to Feller (1937). Statement (b) can be clarified as follows. Essentially, there are two ways in which Feller's condition

$$\lim_{n \to \infty} \max_{k \leq n} \frac{\text{var}(X_k)}{\text{var}(X_1 + X_2 + \cdots + X_n)} = 0$$

may be violated:

(1) $\lim_{n \to \infty} \text{var}(X_1 + X_2 + \cdots + X_n) < \infty$ so that $\lim_{n \to \infty} \max_{k \leq n} \text{var}(X_k)/\text{var}(X_1 + X_2 + \cdots + X_n) > 0$. In this case $\text{var}(X_n)$ approaches 0 with X_n behaving more and more like a constant allowing whatever idiosyncratic patterns distribution of earlier random variables to remain uncorrected. Hence the only way in which $X_1, X_2, X_3, \ldots$ can obey the central limit law is to have all $X_1, X_2, X_3, \ldots$ already normally distributed to begin with (see Problem 6.5.5).

(2) $\lim_{n \to \infty} \text{var}(X_1 + X_2 + \cdots + X_n) = \infty$ but $\text{var}(X_n)$ approaches ∞ so fast that $\lim_{n \to \infty} \max_{k \leq n} \text{var}(X_k)/\text{var}(X_1 + X_2 + \cdots + X_n) > 0$. In this case the idiosyncratic patterns of distribution of later random variables forever predominate, so that the only way in which $X_1, X_2, \ldots$ can obey the central limit law is to have all later random variables from some point on perfectly normally distributed (or become increasingly more perfectly normally distributed).

Before we give a formal proof of Proposition 6.5.2 let us note that since

$$\text{var}_{(\epsilon)}(X) = \int_{|x - \mu_X| < \epsilon} (x - \mu_X)^2 P_X(dx)$$

we have the following lemma.

6.5.1 LEMMA

If $\text{var}(X) > M$, then

$$\frac{\text{var}_{(\epsilon)}(X)}{\text{var}(X)} < \frac{\epsilon^2}{M}$$

In other words, for a fixed ϵ, the larger the variance of X the smaller the ratio of $\text{var}_{(\epsilon)}(X)$ to $\text{var}(X)$, that is, $\text{var}_{(\epsilon)}(X)$ captures a smaller fraction of $\text{var}(X)$.

PROOF OF PROPOSITION 6.5.2

Suppose Feller's condition does not hold, that is, $\max_{k \leq n} \sigma_k^2 / B_n^2 = \max_{k \leq n} \text{var}(Y_{n,k})$ does not approach 0 as n approaches ∞. This means there exists a $\delta > 0$ such that

$$\max_{k \leq n} \text{var}(Y_{n,k}) > \delta$$

for infinitely many values of n. For each such value of n let r denote that value of $k(1 \leq k \leq n)$ for which

$$\text{var}(Y_{n,r}) = \max_{k \leq n} \text{var}(Y_{n,k})$$

Now from $\text{var}(Y_{n,r}) > \delta$ it follows by Lemma 6.5.1 that

$$\text{var}_{(\epsilon)}(Y_{n,r}) < \frac{\epsilon^2}{\delta} \text{var}(Y_{n,r})$$

Hence for $\epsilon^2 < \delta$ $\text{var}_{(\epsilon)}(Y_{n,r})$ will never capture more than a fixed fraction (namely ϵ^2/δ) of $\text{var}(Y_{n,r})$ so that $\text{var}_{(\epsilon)}(Y_{n,r})$ will miss at least a fixed fraction (namely $\theta = 1 - \epsilon^2/\delta$) of $\text{var}(Y_{n,r}) > \delta$. Thus it will be impossible for $\sum_{k=1}^{n} \text{var}_{(\epsilon)}(Y_{n,r})$ to exceed $1 - \theta\delta$, let alone approach 1. In other words Lindeberg's condition cannot hold in the absence of Feller's condition.

We conclude this section by deriving several central limit theorems from that of Lindeberg. A proof of Lindeberg's theorem by means of characteristic functions can be found in Gnedenko's book.

6.5.3 PROPOSITION
Lindeberg's theorem implies Theorem 6.5.1, that is, a sequence of independent random variables $X_1, X_2, X_3, \ldots$ all identically distributed as some X having a finite variance obeys the central limit law.

PROOF
We need only show such a sequence of random variables satisfy Lindeberg's condition. Assuming without loss of generality that $\mu(X) = 0$ and $\sigma(X) = \sigma$, since for each n, $Y_{n,k} = (X_k - \mu_X)/B_n = X_k/\sqrt{n}\sigma$ are all identically distributed as $X/\sqrt{n}\sigma$ for $k = 1, 2, \ldots, n$, we have

$$\sum_{k=1}^{n} \text{var}_{(\epsilon)}(Y_{n,k}) = n \, \text{var}_{(\epsilon)}\left(\frac{X}{\sqrt{n}\sigma}\right)$$

$$= n \int_{|x/\sqrt{n}\sigma| < \epsilon} [x^2/n\sigma^2] P_X(dx)$$

$$= \frac{1}{\sigma^2} \int_{|x| < \sqrt{n}\sigma\epsilon} x^2 P_X(dx)$$

Since the last integral tends to σ^2 as n tends to ∞

$$\lim_{n \to \infty} \sum_{k=1}^{n} \text{var}_{(\epsilon)}(Y_{n,k}) = 1$$

and Lindeberg's condition is satisfied.

Before proceeding further we need to point out an alternative form of Lindeberg's condition.

6.5.2 LEMMA

The following two conditions are equivalent:

$$\lim_{n \to \infty} \frac{1}{B_n^2} \sum_{k=1}^{n} \int_{|x-\mu_k| < \epsilon B_n} (x - \mu_k)^2 P_k(dx) = 1$$

$$\lim_{n \to \infty} \frac{1}{B_n^2} \sum_{k=1}^{n} \int_{|x-\mu_k| \geq \epsilon B_n} (x - \mu_k)^2 P_k(dx) = 0$$

PROOF

In terms of $Y_{n,k} = (X_k - \mu_k)/B_n$ we need only show the equivalence of

$$\lim_{n \to \infty} \sum_{k=1}^{n} \text{var}_{(\epsilon)}(Y_{n,k}) = 1$$

and

$$\lim_{n \to \infty} \sum_{k=1}^{n} \text{var}_{\epsilon}(Y_{n,k}) = 0$$

where we let $\text{var}_{\epsilon}(Y_{n,k}) = \text{var}(Y_{n,k}) - \text{var}_{(\epsilon)}(Y_{n,k})$. This follows immediately from

$$\sum_{k=1}^{n} \text{var}_{\epsilon}(Y_{n,k}) = \sum_{k=1}^{n} [\text{var}(Y_{n,k}) - \text{var}_{(\epsilon)}(Y_{n,k})]$$

$$= \text{var}\left(\sum_{k=1}^{n} Y_{n,k}\right) - \sum_{k=1}^{n} \text{var}_{(\epsilon)}(Y_{n,k})$$

$$= 1 - \sum_{k=1}^{n} \text{var}_{(\epsilon)}(Y_{n,k})$$

6.5.3 THEOREM (Liapounov)

A sequence of independent random variables $X_1, X_2, X_3, \ldots$ all having finite "absolute central moments" of $(2 + \delta)$th order for some fixed $\delta > 0$ such that

$$\lim_{n \to \infty} \frac{1}{B_n^{2+\delta}} \sum_{k=1}^{n} \int_R |x - \mu_k|^{2+\delta} P_X(dx) = 0$$

obeys the central limit law.

PROOF

Rewriting the complicated expression conveniently as

$$\sum_{k=1}^{n} \int_R |(x - \mu_k)/B_n|^{2+\delta} P_X(dx) = \sum_{k=1}^{n} \text{var}^{(\delta)}(Y_{n,k})$$

we shall merely show (cf. Lemma 6.5.2) that

$$\sum_{k=1}^{n} \text{var}^{(\delta)}(Y_{n,k}) \to 0 \quad \text{implies} \quad \sum_{k=1}^{n} \text{var}_{\epsilon}(Y_{n,k}) \to 0$$

Now in general for any $\varepsilon > 0$ and $\delta > 0$ we have (assuming without loss of generality that $\mu_X = 0$)

$$\text{var}^{(\delta)}(X) = \int_R |x|^{2+\delta} P_X(dx) \geq \int_{|x| \geq \epsilon} |x|^2 |x|^\delta P_X(dx)$$

$$\geq \varepsilon^\delta \int_{|x| \geq \epsilon} |x|^2 P_X(dx) = \varepsilon^\delta \text{var}_\epsilon(X)$$

Hence

$$\sum_{k=1}^n \text{var}_\epsilon(Y_{n,k}) = \varepsilon^{-\delta} \sum_{k=1}^n \text{var}^{(\delta)}(Y_{n,k}) \to 0$$

and the theorem is proved.

6.5.2 COROLLARY
A sequence of independent random variables $X_1, X_2, X_3, \ldots$ obeys the central limit law if there exists a constant M such that

$$|X_k - \mu_k| \leq M \qquad \text{for all} \quad k$$

and furthermore $B_n^2 = \text{var}(X_1 + X_2 + \cdots + X_k)$ approaches ∞.

PROOF
In view of Liapounov's theorem it suffices to show that

$$\sum_{k=1}^n \text{var}^{(\delta)}(Y_{n,k}) \to 0$$

Now since $Y_{n,k}$ is bounded by M/B_n, we have for any $\delta > 0$

$$\sum_{k=1}^n \text{var}^{(\delta)}(Y_{n,k}) = \sum_{k=1}^n \int_{|y| \leq M/B_n} |y|^2 |y|^\delta P_{n,k}(dy)$$

$$\leq (M/B_n)^\delta \sum_{k=1}^n \int_R |y|^2 P_{n,k}(dy)$$

$$= (M/B_n)^\delta \sum_{k=1}^n \text{var}(Y_{n,k})$$

$$= (M/B_n)^\delta \to 0$$

as n approaches ∞, and the corollary is proved.

6.5 PROBLEMS

1. Given identically distributed independent random variables $X_1, X_2, X_3, \ldots$ with mean values μ and standard deviations σ, show that

$$S_n = X_1 + X_2 + \cdots X_n \quad \text{and} \quad A_n = \frac{1}{n}(X_1 + X_2 + \cdots + X_n)$$

have identical standardizations and therefore for large values of n, A_n is approximately $(\mu, \sigma/\sqrt{n})$ normally distributed.

2. Monte Carlo method. Suppose a certain "random mechanism" is capable of producing a number in the interval $[a, b]$ completely at random. Let us refer to such a mechanism, which in effect simulates a random variable X having a uniform distribution over $[a, b]$, as a *table of random numbers* $[a, b]$. In view of the weak law of large numbers (Theorem 6.1.1) such a table may be used to estimate an integral $J = \int_a^b f(x)\, dx$ as follows:

$$\frac{J}{b-a} = \int_a^b f(x)\,\frac{1}{b-a}\,dx = E[f(X)]$$

thus

$$J = (b-a)E[f(X)]$$

Hence by the weak law of large numbers

$$(b-a)\frac{f(X_1) + f(X_2) + \cdots + f(X_n)}{n}$$

should be a good approximation of J for large n where $X_1, X_2, \ldots, X_n$ are n "drawings" from the table of random numbers $[a, b]$. Using the central limit theorem (Theorem 6.5.1), we can determine how good such an approximation is.

In approximating the integral $J = \int_0^1 e^{-x^3}$ by the Monte Carlo method with a table of random numbers $[0, 1]$:

   a. If $n = 10,000$ drawings are made to arrive at an approximation, how sure are we that the error is not greater than 0.02? (Use data from page 96.)

   b. If we want to be at least 99.74% sure that the error is not greater than 0.01, how many drawings must we make? Note: $0.9974 = 2[\Phi(3) - 0.5]$. (Cf. page 96.)

3. In estimating the integral $J = \int_0^{\pi/2} \sin x\, dx$ by the Monte Carlo method how many drawings should we make in order to be at least 99.74% sure that the error does not exceed 0.1?

4. In a chicken farm an egg is known to hatch with probability 0.9. Use the Demoivre-Laplace limit theorem to determine how many eggs one must incubate in order to be at least 99.87% sure that 9000 (or more) chickens will be hatched.

5. Given a sequence of independent random variables $X_1, X_2, \ldots$ show the following.
   a. Feller's condition

$$\lim_{n \to \infty} \max_{k \le n} \frac{\text{var}(X_k)}{\text{var}(X_1 + \cdots + X_n)} = 0$$

implies $\lim_{n \to \infty} \text{var}(X_1 + \cdots + X_n) = \infty$.

b. If X_k is normally distributed with mean 0 and variance 2^{-k}, then the sequence obeys the central limit law without satisfying Feller's condition, whence show that Lindeberg's condition is not a necessary condition for the central limit law.

6. Given a sequence of independent random variables $X_0, X_1, X_2, \ldots$ suppose X_0 is uniformly distributed over the interval $[-1, 1]$ and X_k is normally distributed with mean 0 and variance 2^k for $k = 1, 2, 3, \ldots$; show that the sequence obeys the central limit law without satisfying Feller's condition (let alone Lindeberg's condition). Hint: $\max_{k \leq n} \sigma_k^2/B_n^2 = \sigma_n^2/B_n^2 > \frac{1}{2}$. Calculate the characteristic function of Z_n, the standardization of $S_n = X_1 + \cdots + X_n$.

7. Given a sequence of random variables $X_1, X_2, \ldots$ all having finite variances, if $\lim_{k \to \infty} \text{var}(X_k) = \infty$, then according to Lemma 6.5.1, $\lim_{k \to \infty} \text{var}_{(\varepsilon)}(X_k)/\text{var}(X_k) = 0$ for any $\varepsilon > 0$. A proposition dual to this would be: If $\lim_{k \to \infty} \text{var}(X_k) = 0$, then $\lim_{k \to \infty} \text{var}_{(\varepsilon)}(X_k)/\text{var}(X_k) = 1$ for any $\varepsilon > 0$. Construct a simple counterexample to show that this is false. Hint: Try X_k with $P_k(0) = 1 - 1/2^k$ and $P_k(-k) = P(k) = 1/2^{k+1}$.

8. Show by a trivial counterexample that the central limit theorem for uniformly bounded independent random variables (Corollary 6.5.2) does not hold in the absence of the assumption that the variance of the sum B_n^2 tends to ∞.

9. Show that a sequence of independent random variables X_k obeys the central limit law if there exists a constant M such that $|X_k| \leq M$ for all k. Hint: Use Corollary 6.5.2.

TABLES

TABLE 1 SELECTED VALUES OF THE STANDARD NORMAL DISTRIBUTION

$$y = \frac{1}{\sqrt{2\pi}} \int_{0}^{x} e^{-t^2/2}\, dt$$

x	y	x	y
0.0	0.000	1.64	0.450
0.1	0.040	1.7	0.455
0.2	0.079	1.8	0.464
0.3	0.118	1.9	0.471
0.4	0.155	1.96	0.475
0.5	0.191		
0.6	0.226	2.0	0.477
0.7	0.258	2.1	0.482
0.8	0.288	2.2	0.486
0.9	0.316	2.3	0.489
		2.33	0.490
1.0	0.341	2.4	0.492
1.1	0.364	2.5	0.494
1.2	0.385	2.6	0.495
1.28	0.400	2.7	0.496
1.3	0.403	2.8	0.497
1.4	0.419	2.9	0.498
1.5	0.433		
1.6	0.445	3.0	0.499

TABLE 2 SELECTED VALUES OF THE POISSON DISTRIBUTION

$$p(k;\mu) = e^{-\mu}\frac{\mu^k}{k!}$$

μ \ k	0	1	2	3	4	5	6	7	8	9	10
0.1	0.905	0.091	0.005	0.000							
0.2	0.819	0.163	0.016	0.001	0.000						
0.3	0.741	0.222	0.033	0.003	0.000						
0.4	0.670	0.268	0.054	0.007	0.001	0.000					
0.5	0.607	0.303	0.076	0.013	0.002	0.000					
0.6	0.549	0.329	0.099	0.020	0.003	0.000					
0.7	0.497	0.348	0.122	0.028	0.005	0.001	0.000				
0.8	0.449	0.360	0.144	0.038	0.008	0.001	0.000				
0.9	0.407	0.366	0.165	0.049	0.011	0.002	0.000				
1.0	0.368	0.368	0.184	0.061	0.015	0.003	0.001	0.000			
1.5	0.223	0.335	0.251	0.126	0.047	0.014	0.004	0.001	0.000		
2.0	0.135	0.271	0.271	0.180	0.090	0.036	0.012	0.003	0.001	0.000	
2.5	0.082	0.205	0.257	0.214	0.134	0.067	0.028	0.010	0.003	0.001	0.000
3.0	0.050	0.149	0.224	0.224	0.168	0.101	0.050	0.022	0.008	0.003	0.001
3.5	0.030	0.106	0.185	0.216	0.189	0.132	0.077	0.039	0.017	0.007	0.002
4.0	0.018	0.073	0.147	0.195	0.195	0.156	0.104	0.060	0.030	0.013	0.005
4.5	0.011	0.050	0.113	0.169	0.190	0.171	0.128	0.082	0.046	0.023	0.010
5.0	0.007	0.034	0.084	0.140	0.176	0.176	0.146	0.104	0.065	0.036	0.018
6.0	0.003	0.015	0.045	0.089	0.134	0.161	0.161	0.138	0.103	0.069	0.041
7.0	0.001	0.006	0.022	0.052	0.091	0.128	0.150	0.150	0.130	0.101	0.071
8.0	0.000	0.003	0.011	0.029	0.057	0.092	0.122	0.140	0.140	0.124	0.099
9.0	0.000	0.001	0.005	0.015	0.034	0.061	0.091	0.117	0.132	0.132	0.119
10.0	0.000	0.001	0.002	0.008	0.019	0.038	0.063	0.090	0.113	0.125	0.125

MINIBIBLIOGRAPHY

ELEMENTARY LEVEL

Mosteller, F., *Fifty Challenging Problems in Probability*, Addison-Wesley, Reading, Pa. (1965).

Neyman, J., *First Course in Probability Theory and Statistics*, Holt, Reinhart & Winston, New York (1950).

INTERMEDIATE LEVEL

Feller, W., *An Introduction to Probability Theory and Its Applications*, vol. 1, Wiley, New York (1957).

Gnedenko, B. V., *The Theory of Probability*, Chelsea, New York (1967).

Karlin, S., *A First Course in Stochastic Processes*, Academic Press, New York (1966).

Parzen, E., *Modern Probability Theory and Its Applications*, Wiley, New York (1960).

ADVANCED LEVEL

Breiman, L., *Probability*, Addison-Wesley, Reading, Pa. (1968).

Chung, K. L., *A Course in Probability Theory*, Harcourt Brace Jovanovich, New York (1968).

Feller, W., *An Introduction to Probability Theory and Its Applications*, vol. 2, Wiley, New York (1966).

Loève, M., *Probability Theory*, Van Nostrand Reinhold, New York (1963).

Neveu, J., *Mathematical Foundations of the Calculus of Probability*, Holden-Day, San Francisco (1965).

Royden, H. L., *Real Analysis*, Macmillan, New York (1963).

ANSWERS

CHAPTER 1

Section 1.1

1. (a) $\{1, 2, 3, 4, 5, 6\}$
   (b) $\{r, w, b, rw, rb, wb, rwb\}$
   (c) $\{1, 2, 3, 4\}$
2. (a) $\{1, 2, 3, \ldots\}$
   (b) $\{t: t > 0\}$
   (c) $[8, 12] \times [7, 8]$, a rectangle.
   (d) If x and y denote the lengths of the left-end and the right-end pieces, respectively, then $0 \leq x + y \leq 1$; this inequality can be graphed in the x, y-plane as a triangle $S = \{(x, y): 0 \leq x, 0 \leq y, x + y \leq 1\}$.
   (e) $S = \{\langle x_1, x_2, \ldots, x_n\rangle: \sum_i x_i = 1; x_1 \geq x_2 \geq \cdots \geq x_n; n = 1, 2, 3, \ldots\}$

Section 1.2

1. $\{1, 2, 3\}$, a subset of the set of all positive integers.
2. $A = \{(x, y): 0 \leq x + y \leq 1, x \geq 0.5\}$, a triangle.
   $B = \{(x, y): 0 \leq x + y \leq 1, x \geq 0.5, y \leq 0.25\}$, a trapezoid.
3. If $x_3 < 0.1$, then certainly $x_5 < 0.1$. Hence, the first event A implies the second event B; $A \subset B$.
4. No; it only means $A \cup B = S$.
5. $S = \{hh, ht, th, tt\}$. There are 2^3 subsets of S containing the element hh: for example, $\{hh\}$, $\{hh, ht\}$, $\{hh, ht, th\}$.
6. $E = AB$, $D = (A \cup B)E^c$
7. 190
9. E_7; E_2, E_{12}
10. $\frac{1}{2}$

272

Section 1.3

1. $R_3 = H_1H_2H_3H_4^cH_5^c \cup H_1^cH_2^cH_3H_4H_5 \cup H_1^cH_2H_3H_4H_5^c \cup H_1H_2^cH_3^cH_4^cH_5 \cup H_1^cH_2^cH_3H_4H_5 \cup H_1H_2H_3^cH_4^cH_5^c$

2. $S_1 = \{1, 2, 3, 4, 5, 6\}$; let $A_i = \{i\}$, $i = 1, 2, 3, 4, 5, 6$
   $S_2 = \{1, 2, 3, 4, 5, 6\}$; let $B_j = \{j\}$, $j = 1, 2, 3, 4, 5, 6$
   $S = \{(1, 1), (1, 2), (1, 3), \ldots, (6, 5), (6, 6)\}$
   Let $A_i = \{(i, 1), (i, 2), (i, 3), (i, 4), (i, 5), (i, 6)\}$,
       $B_j = \{(1, j), (2, j), (3, j), (4, j), (5, j), (6, j)\}$,
   then $\{(i, j)\} = A_iB_j$ for $i, j = 1, 2, 3, 4, 5, 6$.

3. $S_1 = \{$alike, different$\}$
   $S_2 = \{$heads, tails, mixed$\}$
   $S_3 = \{hh, tt, ht, th\}$

4. $S = \{bb, gg, bg, gb\}$; $P\{bg, gb\} = \frac{2}{4}$
   What is wrong in: $S = \{$boys, girls, mixed$\}$ so $P\{$mixed$\} = \frac{1}{3}$?
   For three children, $P\{$boys$\} = \frac{1}{8}$.

5. 0.325

6. 0.145

Section 1.4

1. $\frac{5}{36}$

2. 0.10

3. $10!/5!3!2! = 2240$

4. 1/216, 7/216, 19/216, 37/216, 61/216, 91/216.

5. $\frac{1}{2}, \frac{1}{4}, \frac{1}{8}$

6. 0.47, 0.44, 0.42

8. 3 gold, 1 silver; 15 gold, 6 silver

Section 1.5

4. $\frac{2}{3}$

5. $1 - \pi/36$

7. $\frac{5}{9}$

8. $\frac{1}{4}$

CHAPTER 2

Section 2.1

1. $\frac{1}{4}, \frac{1}{2}$

2. $\pi/16$, $\pi/16$, 0, $\pi/8$, approximately $\frac{1}{2}$

3. $\frac{2}{3}$, ($\frac{1}{2}$ is a wrong answer obtained without a correct sample set).

4. $\frac{9}{16}, \frac{1}{4}$

5. $\frac{1}{2}$ for both (a) and (b)

Section 2.2

1. $\frac{1}{4}$

2. $\frac{1}{10}$

3. 0.965

5. 0.88. Hint: Let M_i be the event of being moody on the ith day of the week. First calculate $P(M_2)$ from $M_2 = M_2 M_1 \cup M_2 M_1^c$, then calculate $P(M_3)$ likewise.

6. Set $P(b|A) = 0.02$, $P(a|B) = 0.01$; $P(A) = \frac{2}{3}$, $P(B) = \frac{1}{3}$; then $P(A|a) = 196/197$.

7. 0.534

8. 0.999

10. (b) $P(B|D) = P_D(B) = \sum_i P_D(BA_i) = \sum_i P(BA_i|D) = \sum_i P(A_i|D)P(B|A_i D)$

Section 2.3

1. No; compare $P(A^c)$ and $P(A^c|B)$

2. No

3. Independent

4. Yes; no

5. (d) $P(A_r) = C_r^n p^r (1-p)^{n-r}$
   (e) $55/243 \doteq 0.227$

6. 0.0729, 0.0815

7. $1 - (1-q)^n$, q^n

8. $n \geq \log(1-P)/\log(1-p) \doteq 152$

9. 0.28. No, in fact 0.28 is the largest value among $C_r^{10}(0.25)^r(0.75)^{10-r}$ for $r = 0, 1, 2, \ldots, 10$.

10. $q = 1/\sqrt{2} \doteq 0.707$

11. $P(A_{r,s}) = [n!/r! \, s! \, (n-r-s)!] p^r q^s g^{n-r-s}$

Section 2.4

5. (a) Each elementary event $E = E_1^\delta E_2^\delta E_3^\delta \ldots$, where δ is either void or c (for complementation). Being a countable intersection of members in $\mathscr{S}$, E must be in $\mathscr{S}$ [cf. Problem 1.5.12(iv)].
   (b) Since $E \subset E_1^\delta E_2^\delta \cdots E_n^\delta$ for any n, $P(E) \leq P(E_1^\delta E_2^\delta \cdots E_n^\delta) = P(E_1^\delta)P(E_2^\delta) \cdots P(E_n^\delta) \leq b^n$ where $b = \min(p, q) < 1$.

Section 2.5

1. 38%, a restatement of Example 2 of this section.

2. 0.756, 0.729

4. $P(E_1^3|E_3^3) = P(E_1^3)P(E_3^3|E_1^3)/P(E_3^3) = (\frac{1}{2} \cdot \frac{1}{4})/\frac{1}{8} = 1$ from $q_1 = (0, \frac{1}{2}, \frac{1}{2})$ and $q_3 = q_1 M^2 = (\frac{3}{8}, \frac{4}{8}, \frac{1}{8})$.

5. Let $q = (x, y, z)$ and solve

$$\frac{1}{4}x + 0y + \frac{3}{4}z = x$$
$$\frac{1}{4}x + \frac{2}{3}y + 0z = y$$
$$\frac{1}{2}x + \frac{1}{3}y + \frac{1}{4}z = z$$
$$x + y + z = 1$$

to obtain $x = \frac{4}{11}$, $y = \frac{3}{11}$, and $z = \frac{4}{11}$.

6. Let $p_j^{(n)}$ be the probability of being at position j after n units of time, then $p_j^{(n)} \leq (\frac{1}{2})^n$, so $\sum_{n=1}^N p_j^{(n)} \leq N/2^n \to 0$.

7. Let $M = (p_{ij})$ and $v = (v_1, v_2, \ldots, v_k)$. Write out the k simultaneous linear equations by expanding $qM = q$, and show that the k equations are not independent

by adding the left-hand side and the right-hand side of the k equations separately. The resulting identity shows that the k linear equations are not independent, hence there exists at least one nontrivial solution v.

9. (a) By Definition 2.5.5, $\lim_{n \to \infty} qM^n = qM^\infty = q_\infty$ regardless of q; in particular, if $q = (1, 0, 0, \ldots, 0)$, then the first row of M^∞ is seen to be equal to q_∞. Ditto for other rows.

(b) Let $v_i = (0, 0, \ldots, 1, 0, \ldots, 0)$ with 1 occurring at the ith position, then $\lim_{n \to \infty} v_i M^n = v_i M^\infty = i$th row of $M^\infty = p$, say. Let $q = (x_1, x_2, \ldots, x_n) = \sum_i x_i v_i$, then $qM = \sum_i (x_i v_i) M^\infty = (\sum_i x_i) p = p$.

10. (a) M^2 is positive.

(b) Solve $qM = q$ to obtain $q_\infty = (\frac{1}{3}, \frac{1}{3}, \frac{1}{3})$. From Problem 2.5.9(a), all entries of M are $\frac{1}{3}$.

Section 2.6

1. $1 - e^{-0.15} \doteq 0.14$
2. $e^{-(0.05)(10)}(0.5) \doteq 0.30$
3. $1 - \sum_{k=0}^{9} e^{-10} 10^k / k! \doteq 0.54$
4. The smallest K such that $\sum_{k=0}^{K} \alpha_k(1) = \sum_{k=0}^{K} e^{-96} 96^k / k! \geq 0.99$ is 119.
5. 5 counters
6. 9 counters
7. (a) $1 - e^{-0.2} \doteq 0.18$

(b) Same as in (a). Actually, the typist might tend to be less careful if the last mistake was made a long time ago, thus the Poisson probability space may not be a very realistic model here.

CHAPTER 3

Section 3.1

1. X is discrete, Y is continuous.
5. One-point distribution; W is a constant.
6. 0.01
7. 0.32
10. $1 - 1/\sqrt{e}$
14. $Z > a + b$ implies $X + Y > a + b$ implies $X > a$ or $Y > b$, that is, $\{Z > a + b\} \subset \{X + Y > a + b\} \subset \{X > a\} \cup \{Y > b\}$, hence $P\{Z > a + b\} \leq P\{X + Y > a + b\} \leq P\{X > a\} + P\{Y > b\} = \varepsilon + \eta$.

Section 3.2

6. $F_X(x) = (x/\pi)\sqrt{1 - x^2} + (1/\pi)\arcsin x + \frac{1}{2}$ for $0 \leq x \leq 1$
7. $F_X(x) = P(X \leq x) = x^2 \pi / \pi = x^2$ for $0 \leq x \leq 1$ so that $f_X(x) = 2x$. Intuitively, possible; formally, no.
8. $\pi/2 - \frac{1}{2}, \pi/2 + \frac{1}{2}$
9. $[\frac{1}{2}, 1]$ Hint: Let X be the distance from the point 0 to the particle at any moment. Then $F_X(x) = \frac{1}{2}(1 + \sqrt{x} - \sqrt[3]{1 - x})$ and $F_X(\frac{1}{2}) < \frac{1}{2}$.

Section 3.3

4. $\dfrac{1}{2\sqrt{y}}\,[f_X(\sqrt{y})+f_X(-\sqrt{y})]$

Section 3.4

1. The discrete distribution of $(\check{X},\ \check{Y})$ approximates the continuous distribution of $(X,\ Y)$.
2. X has a continuous distribution without a density function, that is, a singular continuous distribution.
3. Yes for (b) and (d), no for (a) and (c).
4. $\frac{5}{9}$
5. $\frac{1}{6}$
6. Hint: Proceed step by step as follows: (i) If $A \in \mathscr{B}^1$, call $A \times R$ a Borel cylinder. If I is an interval in R^1, call $I \times R$ an interval cylinder, and show that the class of all Borel cylinders is the smallest sigma algebra containing the class of all interval cylinders, recalling that $\mathscr{B}^1$ is the smallest sigma algebra containing all intervals. (ii) Show that $\mathscr{B}^2$ contains the class of all Borel cylinders of the form $A \times R$ by combining the result in (i) and the fact $\mathscr{B}^2$ is a sigma algebra containing the class of all interval cylinders of the form $I \times R$. Show likewise that $\mathscr{B}^2$ contains the class of all Borel cylinders of the form $R \times B$ where $B \in \mathscr{B}^1$. Finally, note that $A \times B = (A \times R) \cap (R \times B)$ and that $A \times R$ and $R \times B$ are already members of $\mathscr{B}^2$.

Section 3.5

1. Enumerate all x (and there can be only finitely many such) in A such that $P_X(x) \geqq \frac{1}{2}$, then all x such that $\frac{1}{2} > P_X(x) \geqq \frac{1}{4}$, then $\frac{1}{4} > P_X(x) \geqq \frac{1}{8}$, and so on.
2. The two-dimensional random vector X has a mixed distribution, partly discrete with $\frac{3}{4}$ of probability mass divided equally among the 60 second points, and partly continuous (singular, however) with $\frac{1}{4}$ of mass uniformly distributed along the circle passing the second points.
7. (b) Monotonicity of F can be expressed as $F(x_2,\ y_2) - F(x_2,\ y_1) \geqq F(x_1,\ y_2) - F(x_1, y_1)$. Keeping x_2, y_1, y_2 fixed while x_1 goes to $-\infty$, we have by (a) $F(x_2, y_2) - F(x_2, y_1) \geqq 0 - 0 = 0$, that is, F is monotone with respect to y.

Section 3.6

1. $\frac{3}{4}$; $F_Z(z)$ rises from $z = -\sqrt{2}$ to $z = \sqrt{2}$.
2. $F_Z(z)$ rises concave-upward from $F_Z(0) = 0$ to $F_Z(1^-) = \frac{1}{4}$, continues rising concave-downward from $F_Z(1) = \frac{1}{2}$ to $F_Z(2^-) = \frac{3}{4}$, and finally $F_Z(2) = 1$.
4. (b) Obtain G_{X+Y} from $G_X(t)$ and $G_Y(t)$, and set it equal to $\frac{1}{11}(t^2 + t^3 + \cdots + t^{12})$. Cancel out t^2 from both sides and derive a contradiction by pointing out that while one polynomial admits a real root the other polynomial does not.
8. $P_X * P_{Y_n}$ has a "multitriangular" distribution, a continuous distribution, tending to P_X, a discrete distribution.
11. $u = x + y + 1,\ v = -x + y + 1$

12. A uniform distribution over a triangle.

14. $f_{U, V}(u, v) = f_{X, Y}(u - a, v - b)$

15. $f_{U, V}(u, v) = f_{X, Y}[(u - a)/c, (v - b)/d]/|cd|$

CHAPTER 4

Section 4.1

1. $1/p$
2. Undefined
3. No
4. Yes
5. Yes
8. $E(X)$ for betting on triples is $29\frac{1}{6}¢$; $E(Y)$ for betting on aces is $35¢$. Hence it is better to bet on the aces.
9. 4.5
10. (a) No, the chance is worth taking.
    (b) No. If one pays $E(X) = \$100$, this amount will not be recovered until the seventh toss; the probability of this is only $(\frac{1}{2})^7$.
11. It is better to be player A for large N. Although there are more even numbers than odd numbers in the roulette wheel, A will not go broke so easily if N is large; meanwhile A can halt the game while he is ahead. If $N = 1$, then certainly it is better to be player B.

Section 4.2

5. True for $n = 3$, false for $n = 4$.

Section 4.3

1. $1/\lambda$, $1/\lambda^2$
2. $(a + b)/2$
3. $1/p$, q/p^2
9. $1/\lambda^2$, r/λ^2
10. rq/p^2

Section 4.4

1. (a) $\mu(X)$ is the center of mass distributed according to P_X.
   (b) The matrix (σ_{ij}) has nonnegative entries, and is symmetric with respect to the diagonal. The diagonal consists of the n marginal variances. The matrix $(\sigma_{ij}) = \Gamma$ is a "nonnegative definite" matrix, meaning that for any row vector $x = (x_1, x_2, \ldots, x_n)$ we have $x\Gamma x^t \geq 0$ where x^t is the transpose of x. This is shown by identifying $x\Gamma x^t$ with the variance (hence nonnegative) of the random variable $\sum_i x_i X_i$.
2. Recall $\text{cov}(X, Y) = \sigma_X \cdot \sigma_Y \rho(X, Y)$
5. $3\sqrt{5/7} \doteq 0.96$
6. Assume without loss of generality $E(X) = E(Y) = 0$. From $E(XY) = p_1 x_1 y_1 + p_2 x_2 y_2 = 0$, deduce, say, $y_1/y_2 = -p_2 x_2/p_1 x_1$, then note $p_2 x_2/p_1 x_1 = 1$ since $p_1 x_1 + p_2 x_2 = E(X) = 0$, and hence $y_1 = y_2$, and so on.

CHAPTER 5

Section 5.1

1. $Y = 1, 2, 3$ with probabilities $\frac{5}{7}, \frac{1}{7}, \frac{1}{7}$.
   $X = 2, 4$ with probabilities $\frac{2}{5}, \frac{3}{5}$.
2. $F_{Y|X=[0, 1/n]}(y) = [1 + ny^3]/(1 + n)y \to y^2$ for $0 \leq y \leq 1$.
4. Equal distribution on four points $(\pm 1, \pm 1)$.
5. $\alpha = \frac{2}{9}, \beta = \frac{1}{9}$
6. Only (a)
7. $\frac{2}{3}$
9. $1 - e^{-1}$.

Section 5.2

1. The graph of f_X is a triangle with the base on the interval $[-3, 1]$ and the vertex at the point $(0, \frac{1}{2})$.
4. No
5. Absolutely continuous and singularly continuous, but not discrete, examples can be found.
6. $f_{Y|x}(y) = \frac{1}{2}y + \frac{7}{4}$ for $0 \leq y \leq 1$.
7. (a) $f_{Y|x}(y) = (xy + 2.25x^2)/(0.5x + 2.25x^2)$
   (b) $P_{Y|X\in(1/2, 9/16)}$ is absolutely continuous with a density function approximately equal to $f_{Y|x}$ for $x = \frac{1}{2}$, hence $P(\frac{1}{2} \leq Y \leq 1 | \frac{1}{2} \leq X \leq \frac{9}{16}) \doteq \frac{15}{26}$.

Section 5.3

2. $(3x + 4)/6(x + 1)$, $(3x^2 + 6x + 2)/36(x + 1)^2$
3. $(n - k)p/(p + q)$, $(n - k)pq/(p + q)^2$
4. (a) $\operatorname{cov}(X, Y) = 1.25$, $\rho = 0.5$
   (b) $\beta = 20$, $\alpha = 30$
   (c) $\hat{Y} = 140$
5. $\mu_V = 0$, $\sigma_V = 1$; $\mu_W = 0$, $\sigma_W = \sqrt{\beta^2 + 1}$, $\rho = \beta/\sqrt{\beta^2 + 1}$
6. $\gamma = \pm \sqrt{3}/2$ $\rho = \frac{1}{2}$

CHAPTER 6

Section 6.1

1. 0.89
9. Since f is continuous, given an ε-neighborhood N of $f(\mu)$, we can find an η-neighborhood H of μ such that $f(H) \subset N$. By Khintchine's law of large numbers, for any arbitrarily small $\delta > 0$, $P(|A_n - \mu| \geq \eta) < \delta$ as long as n is sufficiently large. Consequently, for sufficiently large n we have $P(|f(A_n) - f(\mu)| \geq \varepsilon) < \delta$. Hence,

$$|E[f(A_n)] - f(\mu)| \leq E|f(A_n) - f(\mu)|$$
$$\leq \varepsilon P(|f(A_n) - f(\mu)| < \varepsilon) + M \cdot P(|f(A_n) - f(\mu)| \geq \varepsilon)$$

where the constant M is guaranteed by the boundedness of f. Hence, $|E[f(A_n)] - f(\mu)| \leq \varepsilon \cdot 1 + M \cdot \delta$. Since ε and δ are arbitrary, the proof is complete.

10. From $A_n = (1/n) (X_1 + X_2 + \cdots + X_n)$ with $X_1 + X_2 + \cdots + X_n$ having the (n,p) binomial distribution we see that the random variable $f(A_n)$ has the range $f(0/n)$, $f(1/n), \ldots, f(k/n), \ldots, f(n/n)$ with the corresponding probabilities $C_k^n p^k (1-p)^{n-k}$ for $k = 0, 1, 2, \ldots, n$. Consequently $E[f(A_n)] = \sum_{k=0}^{n} f(k/n) C_k^n p^k (1-p)^{n-k} = B_n(p) \to f(p)$ by Problem 6.1.9. The polynomial $B_n(p)$ is called the Bernstein polynomial.

Section 6.2

3. (b) It suffices to show that for each n, $P(D_1 \cup D_2 \cup \cdots \cup D_n) = 0$, which is equivalent to showing that for each n, $P[(\bigcup_{m=0}^{\infty} D_{n+m})^c] = 0$. But $[\bigcup_{m=0}^{\infty} D_{n+m}]^c = \bigcap_{m=0}^{\infty} D_{n+m}^c$ so that we need only work on $P[\bigcap_{m=0}^{\infty} D_{n+m}^c] = \prod_{m=0}^{\infty} P(D_{n+m}^c) = \lim_{M \to \infty} \prod_{m=0}^{M} P(D_{n+m}^c)$ in view of independence of the D's. Now $\prod_{m=0}^{M} P(D_{n+m}^c) = \prod_{m=0}^{M} [1 - P(D_{n+m})] \le \prod_{m=0}^{M} e^{-P(D_{n+m})} = \exp[-\sum_{m=0}^{M} P(D_{n+m})] \to 0$ since $1 - x < e^{-x} = 1 - x + x^2/2! - x^3/3! + \ldots$ for any $x > 0$, and $\sum_{m=0}^{M} P(D_{n+m}) \to \infty$ because $\sum_{n=0}^{\infty} P(D_n)$ is divergent.

4. (b) $S - E_\varepsilon$ is the event that $|A_n - \mu| \ge \varepsilon$ for at most finitely many n. $P(S - E_\varepsilon) = 1$ means then $\limsup |A_n = \mu| < \varepsilon$ almost surely (i.e., with probability 1).

(c) If $\limsup |A_n - \mu| < \varepsilon$ with probability 1 for any $\varepsilon > 0$, then in particular, $\limsup |A_n - \mu| < 1/k$ with probability 1 for every $k = 1, 2, 3, \ldots$, the intersection of events all having probability 1 still has probability 1.

Section 6.3

1. $p_2 q_1 + p_3 q_1$, which is the coefficient of e^t in $M_{X+Y}(t)$.
3. $M_X(t)$ is defined for $t < \lambda$.
8. No. The A_n have distributions identical to that of X for all n.
9. $\phi_{ax+b}(t) = \exp[i(\mu a + b)t - \frac{1}{2}(a\sigma)^2 t^2]$
10. Not unless they have the same parameter p.
11. (a) $\phi_{-X}(t) = \int e^{it(-x)} P_X(dx) = \bar{\phi}_X(t)$
    (b) $\phi_X(t) = \phi_{-X}(t) = \bar{\phi}_X(t)$, hence ϕ_X must be real-valued.
    (c) Yes.
12. (a) $\phi_X(-t) = \int e^{i(-t)x} P_X(dx) = \bar{\phi}_X(t)$
    (b) $\phi_X(-t) = \bar{\phi}_X(t)$ by (a), but if ϕ_X is real-valued, then $\bar{\phi}_X(t) = \phi_X(t)$.

Section 6.4

3. No
4. No
5. No
6. One-point distribution at 0; uniform distribution over $[-1, 1]$.

Section 6.5

1. Since $\mu(S_n) = n\mu$, $\sigma(S_n) = \sqrt{n}\,\mu$, $\mu(A_n) = \mu$ and $\sigma(A_n) = \sigma/\sqrt{n}$, the standardizations of S_n and A_n are respectively $[X_1 + \cdots + X_n - n\mu]/\sqrt{n}\,\sigma$ and $[(1/n)(X_1 + \cdots + X_n) - \mu]/[\sigma/\sqrt{n}]$, which are clearly identical. Hence by Theorem 6.5.1, for large values of n, A_n is approximately normally distributed.

2. (a) At least 95% sure
   (b) 30,000 drawings
3. 210 drawings
4. 10,101 eggs
8. Let X_1 be uniformly distributed over, say, [0, 1], and let X_2, X_3, ... be all constant random variables.

INDEX

72 73 74 75 76 9 8 7 6 5 4 3 2 1